The Book of *Honu*

D1504026

The Book of

Peter Bennett and
Ursula Keuper-Bennett

A Latitude 20 Book
University of Hawai'i Press | Honolulu

HONU

Enjoying and
Learning about
Hawai'i's
Sea Turtles

To Clothahump

. . . .

Clothahump, the first sea turtle
we ever met, known at our dive
site from 1988 to 1993. Sketch in
watercolor pencil on illustration
board, 10" x 15". Ursula Keuper-
Bennett, winter 2003.

© 2008 University of Hawai'i Press
All rights reserved
Printed in China

14 13 12 11 10 09 08   6 5 4 3 2 1

Library of Congress Cataloging-in-Publication Data
Bennett, Peter, 1947-
   The book of honu : enjoying and learning about
Hawai'i's sea turtles / Peter Bennett and Ursula
Keuper-Bennett.
     p. cm.
"A Latitude 20 Book."
Includes bibliographical references.
ISBN 978-0-8248-3127-1 (pbk. : alk. paper)
   1. Green turtle — Hawaii.   I. Keuper-Bennett, Ursula,
1949–   II. Title.
QL666.C536B46 2008
597.92'809969 — dc22

                                        2008008660

University of Hawai'i Press books are printed on acid-
free paper and meet the guidelines for permanence and
durability of the Council on Library Resources.

Designed by April Leidig-Higgins

Printed by Everbest Co., Ltd.

Contents

Acknowledgments

This book would not exist without the help and encouragement of numerous people. First among these is our mentor and good friend George H. Balazs, who taught us not only about turtles but also how to approach a subject (not just turtles, but any subject) in a thoughtful, rigorous, and scientific manner.

We are equally indebted to Jose Danobeitia, one of the original founders of MVS Solutions Inc. and president since its inception. His support for our efforts has never flagged, and without it our turtle experiences simply would not have been possible.

We also must thank (in no particular order) Eve Clute, Skippy Hau, Glynnis Nakai, Tim West, Randy Miller, Mickey McAfee, John Gorman, Kalei Tsuha, all the owners of Captain Nemo's/Pacific Dive over the years, the numerous beach people of Kamehameha Iki Park (who have helped protect 5690's nests and hatchlings), and especially Blue Robinson of the Nohonani and its resident managers, George Kragca and Pete Macdonald (both deceased), Pat Cerretani, and Bill Lentz. Special thanks to Wayne and Margot. Thanks to our editors Keith Leber and Lee S. Motteler, and to the University of Hawai'i Press for taking on this project.

Finally, thanks to everyone who cares about the turtles. They need all the friends they can get.

1 How It All Began

. . . .

A promise is a promise, Lieutenant Dan.
— Forrest Gump

A promise kept

On a beautiful July day in 1988, a hundred yards from the West Maui shore and fifteen feet underwater, we had just finished photographing an eel. We looked up and there she was — the cutest thing we'd ever seen in any ocean. A sea turtle stole our hearts.

While we don't know how she felt, we do know that upon seeing the little turtle, we were instantly infatuated. The next day, we saw her again. She was soon a regular highlight of our dives and lives. We eventually named her Clothahump, after the turtle wizard in Alan Dean Foster's *Spellsinger* books, because that was the only turtle name we knew. Although we never learned whether she was truly a female, somehow we believed that she was.

She was exceptionally friendly. The following summer, she began coming to swim with us the moment we arrived at her reef. Over the next year, she grew quickly. We were happy to see her flourish. There were hints of trouble, however.

We'd met more turtles, but some of them had tumors. By 1991, the number of afflicted turtles was rising. In 1992, we noticed some suspicious white spots on Clothahump's neck and shoulders. This worried us.

We saw her only once in 1993. The white spots had mushroomed into tumors. We both cried underwater. That day, we vowed to tell Clothahump's story and somehow help the ocean community on her reef. We believed that the turtles were seriously threatened, and we promised to do whatever we could to save them. This book is a direct result of that pledge.

Finding answers

We began as two sport divers with an interest in underwater photography. We weren't marine biologists and had no background in conservation. We knew nothing about the Hawaiian green turtle, or *honu* as the Hawaiians call them.

Our encounter with Clothahump eventually led to more than twenty-five hundred dives with *honu*. We accumulated hundreds of hours of videotape, thousands of photos, and a database containing notes on over 750 *honu*. That's the longest and most detailed record ever made of a group of sea turtles in their underwater habitat.

Once we were interested in *honu,* we naturally became curious about them. We discovered that we could find ready answers to some of our questions: How long can they stay underwater? Where are they hatched? What do they eat? What threats do they face?

Other questions turned out to be difficult because the answers simply aren't known yet and perhaps never will be. For example: How long do they live? How do they navigate during their migrations? Where on the high seas do they spend the first years of their lives? How do they survive out there? No one really knows.

We read what we could find, but aside from scientific papers, there was almost nothing about *honu.* By watching and documenting the turtles, we gradually learned the answers to some questions ourselves. We knew we had a lot more to learn, however, and the best way to do that was to consult the professionals.

In 1993, after we'd been watching *honu* at Honokōwai, Maui, for a few years, we attended the Thirteenth Annual Sea Turtle Symposium on Jekyll Island, Georgia. We learned — and continue to learn — much from the friends we made there and at subsequent symposia; however, our primary purpose for attending had an unexpected outcome.

We were eager to meet the experts because we thought they'd be able to give us insight into what we were seeing. To our surprise, once the turtle specialists heard that we'd been observing turtles underwater for years, they began asking *us* questions: How did we tell them apart, whether they got along together, if they had favorite spots, just what is it that they did all day long — the same sort of queries that everyone had.

We soon realized that there had been little underwater observation of sea turtles. In fact, we learned that almost everywhere outside of Hawai'i, it wasn't even a practical idea. It dawned on us that we had to find the answers ourselves.

One of our earliest photos of Raphael, a *honu* we met in the summer of 1992, before we ever attended a Sea Turtle symposium. She's been a resident at Honokōwai ever since. We didn't discover that she was a female until 2004, when she showed up with an identifying mark that had been placed on her shell while she was nesting.

This cut both ways. It was frustrating not to have the information we wanted. Yet it was exhilarating to find out that we were learning things that no one else appeared to know — that we were discoverers. With that perception, we also understood that we now had the responsibility of making our observations available to anyone as curious as we were.

You are reading our attempt to live up to that obligation. Our goal is to give you the book we wanted so badly in 1988. The organization reflects our own path of learning about *honu:* First we wanted to know where to find them, then we got interested in their biology, which led to investigating their life cycle and behavior. A scientist would approach the subject differently, but we thought this sequence would better suit a budding turtle enthusiast.

Although we're writing about the Hawaiian green turtle, much of the information also applies to sea turtles in general. When there are aspects that apply only to the Hawaiian green turtle, we'll point them out. We'll provide answers whenever we can and try to shed some light on the mysteries.

We'll also share our own opinions and insights, starting with this: Sea turtles live a long time. Perhaps the most important thing we've learned from *honu* is persistence. In order to even begin to grasp the nature of the sea turtle, it is imperative to keep studying them for a long, long time.

That's one reason it's taken us seventeen years to get down to writing this book.

2 So You Want to See a Turtle

. . . .

I've always maintained that the *honu*
are their own best ambassadors.
— George H. Balazs

Best place in the world to see sea turtles

If you want to see marine turtles in their natural habitat, Hawai'i is the best place on the planet. That's because for sea turtles, Hawai'i is by far the best place to be! No other turtle population is blessed with such a lucky combination of isolated geography, effective protection, and year-round Spirit of Aloha.

Hawai'i's turtles are 100 percent Hawaiian, living their entire lives within the island chain and surrounding ocean waters. That makes it easier to keep them from harm. Other sea turtles grow up in the waters of one country, then migrate through the waters of other countries to lay eggs in the sands of yet another. Rarely do all of the involved countries effectively protect the turtles.

Here again, Hawai'i's turtles have been most fortunate. State law protected them from commercial harvesting in 1974, four years before the U.S. Endangered Species Act gave them full protection. Unlike in many other parts of the world, Hawai'i and the United States actually enforce protection laws.

Hawai'i's people have not only provided their turtles with safe harbor but have consistently come to the aid of any *honu* in distress. This has produced an animal that goes about its business under the full gaze of residents and vacationers alike. Few creatures can summon the confidence that many *honu* routinely exhibit around humans. With each new turtle encounter, another human is charmed and *honu* can chalk up one more turtle fan.

George Balazs is leader of Marine Turtle Research for the NOAA Pacific Islands Fisheries Science Center, Protected Species Division, located in Hawai‘i. He's Hawai‘i's acknowledged turtle expert. He points out that *honu* are right up there with Hawai‘i's humpback whales in terms of marine life popularity. The turtles are much more common, easier to see close up in safety, and available all year round.

The people of Hawai‘i clearly love their turtles. Evidence of this abounds: *honu* shirts, *honu* hats, *honu* stickers, *honu* necklaces, *honu* earrings, *honu* wind chimes, *honu* tattoos — a celebration of *honu* everywhere, and why not? *Honu* are beautiful and graceful — charisma in a shell, and above all, nature's work of art.

Spotting *honu* without getting wet

If you've seen a *honu,* then you know the delight that first contact brings. If you haven't, we hope to prepare you for it and bring that precious moment closer. Plentiful and tolerant turtles combined with Hawai‘i's inviting waters mean that *anyone* can experience the thrill of meeting *honu.*

Honu in captivity

On O‘ahu, Sea Life Park has provided a sanctuary for nearly two dozen breeding *honu* for many years. These turtles were obtained in 1968, before laws protecting them had even been conceived. An exception in the legislation made it legal to keep the captive *honu*. Since 1976, the turtles have been nesting successfully on Sea Life Park's artificial beach, so if you visit at the right time you'll also get to see *honu* hatchlings.

The Maui Ocean Center maintains a saltwater pool with an underwater viewing area. It features several young *honu* obtained from Sea Life Park. A year old upon arrival, they spend about two years on display before their release into the wild. Meanwhile, they are stars of one of the center's most popular exhibits, making new friends for the *honu* every day.

Little *honu* also charm guests at the Mauna Lani Bay Hotel on the Big Island. There, a series of beautiful interconnected pools hold several tiny turtles originally from Sea Life Park. When they are large enough, the hotel releases them during special Fourth of July ceremonies known as Turtle Independence Day, a custom that began in 1989.

Honu in the wild

Of course, even seeing *honu* up close in a beautiful aquarium setting is only second best compared to seeing them swimming and living in their

Photo of Amuala, taken in 1997 before we learned that he was a male. We first met Amuala in 1995, when he was small enough for us to classify him as a juvenile. He was tolerant from the beginning, making him a perfect model for photographs.

natural habitat. In Hawai'i, all you need to see a sea turtle in the wild are some hints, combined with a little patience.

Before we start, however, let's review some responsible viewing guidelines that the National Marine Fisheries Service and the State of Hawai'i recommend. Note that these do *not* replace federal or state law:

- Always keep a safe distance. Please do not chase, closely approach, surround, swim with, or attempt to touch marine wildlife.
- For sea turtles, please remember that feeding, touching, or attempting to ride them can cause distress. Please observe from a distance and allow them an escape route to deeper water.
- Never entice marine wildlife to approach you.
- Be careful not to surprise marine wildlife.

We'll emphasize and expand upon these points throughout the book. Now, back to finding turtles in the wild. The first thing to learn is

Maui Ocean Center is a terrific place to learn about Hawai'i's sea creatures. The undeniably cute young *honu* in their sea turtle tank attract plenty of visitors. As interesting as it was to compare MOC *honu* behavior with what we see in the wild, it was even more fascinating to study how people watched *honu.* Kids were the most creative when it came to observing turtles. For example, they'd flip open a cellphone, aim at a turtle, and *click!*—a friend on the mainland would have an instant picture of a Hawaiian shelled wonder. What's almost as good as watching *honu?* Watching kids watch *honu.*

where to look. The key here is that *honu* are primarily marine vegetarians. They feed on seaweeds—*limu* in Hawaiian—that typically grow close to shore. They especially like the red seaweed known as *Pterocladiella capillacea* to scientists and *loloa* to Hawaiians. At low tide, you can see this alga growing on the rocks in many places. Almost anywhere this *limu* grows in abundance is a prime location for spotting turtles.

Next, you need to know when to look. In some areas you can see *honu* feeding at almost any time of the day, but in most places the best times to watch for turtles are late afternoons when the sun rides low and just before dawn when there is barely enough light to see.

Look for turtle heads. Sea turtles breathe air, mainly through their mouths. When they're feeding, they typically come up to breathe every five minutes or so. Keep watching the water around a place where you know *loloa* and other *limu* grow. You are almost certain to see a "popper," our term for a turtle head popping up for a quick breath before ducking back down for more food. *Honu* have become so common that you are often rewarded with a popper just by keeping a watchful eye on almost any stretch of water close to shore.

These photos show the favorite food of the Honokōwai *honu,* the *limu Pterocladiella.* From top to bottom: *Pterocladiella* exposed at low tide, submerged at high tide, and being consumed at *honu* suppertime. All of these pictures were taken right at the shoreline.

Some places attract more turtles than others, probably because the food they like grows there. At the right time at such a place, it's common to see two, three, or more *honu* heads up at the same time, often as close as the water's edge.

Don't assume that two or three heads means only two or three *honu* are about. For every head you see break the surface, there are probably three or four turtles under the water, actively feeding.

Sometimes you can find a popular foraging spot in a place you wouldn't expect. For example, every evening starting about an hour before sunset, turtles gather to feed along the shore just north of the Māla Wharf in Lahaina, Maui. On the *mauka* (inland) side, constant traffic flows along Front Street, while the *makai* (sea) side is busy with boats coming and going from the wharf and adjacent anchorage.

None of this bothers *honu*. They are simply interested in the *Pterocladiella* that grows in the shallows. Fortunately for turtle watchers, there are a few places to park right by the side of the road. Because of the timing, you can enjoy not only the spectacle of numerous *honu* heads popping up to breathe but also one of West Maui's famous sunsets.

Sunset and dawn aren't the only times to watch for *honu*. You can also spot turtles during the day, but usually only when the sun is behind you. Otherwise, the contrast is too high.

One of Hawai'i's great charms is that since the remarkable recovery of the *honu* in recent years, any stretch of coastline will do. Watch the water carefully. The chances that you'll be rewarded are good, but you will need patience. *Honu* don't breathe as often when they're resting on the bottom.

During the day, many *honu* like to lie about on the reef, typically within a couple of hundred yards from shore. To spot these turtles, start scanning at the shoreline and work out toward deeper water.

You begin at the water's edge because more and more turtles have started to feed during daylight, so there's some chance you'll see a popper close to shore. As you look farther out, start watching for the backs of brownish shells. These are *honu,* up for a breath.

When resting turtles surface to breathe, their technique differs from when they are feeding. Instead of popping up quickly, *honu* float at the surface for a while, usually with head down so that they can watch below. Several times during this interlude, they breathe: The head goes up, there is a whoosh and a gasp as air is taken in — and though the head is raised for only a moment, that's when they are easiest to spot.

Without question, the best time to observe and photograph turtles is on those special days when the air is still, the sun is shining bright, and the ocean is millpond calm. When *honu* go up for air in such splendid conditions, they will linger on the surface, flippers spread for balance, and have a long look around. The similarity in appearance to large soaring birds is most pronounced at times like these.

The throat of a *honu* is creamy white, or nearly so. As the head rises, the glistening wet throat expands to help take in air. It flashes in the brilliant sun, making it readily visible from the shore. This is your tip-off. Once you've spotted a turtle, scan the nearby water for others. Usually, where there's one turtle you'll find more.

You can use binoculars, but remember that doing so limits your field of vision. Until you've located the turtles, scan at least some of the time with the naked eye. That gives your peripheral vision a chance to pick up a throat-flash. You might not spot *honu* immediately, but you will know which way to point your field glasses.

There are lots of places where *honu* congregate close enough to shore that you can catch sight of them. For example, you usually can see turtles

A *honu* we know only as 1999 Turtle 153 with head raised for a breath, providing a perfect example of the way a turtle's white throat reflects light almost as well as a mirror. This makes it possible for you to spot a turtle from shore, even at a distance of a couple hundred yards.

any time of day at the south end of Kaloko-Honokōhau National Historical Park, just north of Kailua-Kona on the Big Island. Enter through the Honokōhau Small Boat Harbor and take the park trail to the right. Almost immediately you'll come upon Honokōhau Beach. Usually, you'll see two or three poppers within moments.

Honu on the beach

While you're at Honokōhau Beach, look carefully up and down the shore. You just might see a turtle lying in the sand. In some parts of Hawai'i, *honu* have become so unafraid of humans that they are changing their behavior. One of these changes is where and how often they bask on the beach or rocks.

Honu have long been known to crawl up onto shore to bask in the remote Northwestern Hawaiian Islands. This chain of islets and shoals extends northwest to Midway and Kure Atolls, hundreds of miles from Honolulu. The U.S. government officially protected the area in 1909, and access has long been strictly controlled. In 2006, the chain was declared a national monument and renamed the Papahānaumokuākea Marine National Monument, extending even greater protection.

After flipping sand back over the shell to help keep cool, a *honu* dozes off at Kaloko-Honokōhau National Historical Park on the Big Island, unconcerned about the humans also using the beach. From a distance, it would be easy to overlook this turtle.

Because these islands are almost completely uninhabited, the turtles did not have to fear human presence. Now, it seems *honu* have realized that it is safe to come ashore even where humans abound. There are increasing reports from around the main Hawaiian Islands that turtles are sunbathing without concern. There could be a place like that near you.

Even busy Oʻahu has such a site on its famous North Shore: Laniākea, or as George Balazs likes to call it, "HonuLani, Home of the Heavenly Turtles." Since it has now become more widely known, however, you are more likely to hear it called by the inevitable nickname of "Turtle Beach."

Located between Haleʻiwa Beach County Park and Waimea Bay along the Kamehameha Highway, Laniākea is a magical place where turtles and humans interact daily. *Honu* are making human friends while providing a thrill that cannot be matched anywhere else in the world. If recent trends continue, the turtles at other basking spots will become more tolerant, providing increased opportunities for people to get to know these charming creatures.

If you are lucky enough to learn of a turtle-basking beach, remember that *honu* are still wild animals, protected under both Hawaiʻi State and U.S. federal law. You must respect the turtles at all times. Please show them your aloha. Any action that prompts a *honu* to retreat into the water, such as touching, could be considered illegal. More important, though, is to see it from the turtle's point of view: You've ruined a nice sunbath. How rude!

In some places, like Honokōhau and Laniākea, the turtles feed right at the waterline even when there are people in the water next to them. They've completely lost their fear of humans. Again, under these circumstances you must not forget to show your respect for the *honu*.

Remember, the turtles are big, their shells are hard, they are in their natural element while you're not, and they aren't looking out for people. *Honu* have large and powerful flippers, so when a turtle is trying to move around in the shorebreak to nibble a tasty morsel, you don't want to get in the way. The most likely accident, however, would be an untimely wave that tosses the turtle right at you. If you are struck by the edge of the turtle's shell, the *honu* won't mind — but you definitely will.

You and the *honu* will be safe and happy as long as you use common sense and remember Rule No. 1: Respect *honu* — show them aloha.

Snorkeling with *honu*

Seeing a *honu* basking or feeding right at the shoreline in broad daylight is certainly a wonderful experience. There's no doubt, however, that the best place to see turtles is in their deeper natural environment: under the water, where they live almost all of their lives. There you can see for yourself the grace and beauty of *honu* in motion — and you must. Any description we put on paper would be inadequate.

Snorkeling is the simplest way to have such an experience. The equipment is cheap to rent and almost anyone can use it. If you don't know how to snorkel, seeing a *honu* is the best reason there is to learn.

You'll need to know where to snorkel. Most large turtles tend to stay farther from shore and in water too deep or dangerous for most snorkel-

Most people are satisfied to watch the baskers on the sand right at Laniākea, but if you stroll to the left along the shoreline for a few minutes, you are likely to see *honu* feeding right in the surf at the water's edge. These turtles are so determined to get at the *limu* growing in the nooks and crannies of the rocks that they are oblivious to the waves that break on top of them and toss them around.

ers, but fortunately there are exceptions. At some locations, you can find large *honu* within easy snorkeling range.

For example, on O'ahu you can snorkel over the reef and see big *honu* right in front of the Sheraton Hotel off Waikīkī Beach, and snorkelers commonly see turtles at Hanauma Bay. On Maui, you can usually find several large *honu* around Black Rock, a well-known snorkeling location in Kā'anapali. These are three of the busiest beaches in Hawai'i.

Ask for the local *honu* spot at the place where you rent your equipment or at the local dive shop. There's sure to be at least one. While it might be overcrowded, you can be certain of one thing: Because so many people

Pi'i (Hawaiian for "to mount"—don't ask) hangs at the surface between breaths. An unidentified male *honu* makes the trip to the surface in the background. This turtle encounter required a kayak, since these *honu* are too far out for safe snorkeling from shore.

know about these places, the *honu* you'll find there are extremely tolerant. If they weren't, they wouldn't stay there.

Without a doubt, the unqualified best place to snorkel with *honu* is Laniākea. One caution: Make sure conditions are safe for snorkeling. The waves here can get big and dangerous, especially in winter. Remember, the North Shore of O'ahu—and, in fact, much of the coastline of the Hawaiian Islands—has a reputation for high surf.

If conditions are safe, leave the turtles in the shallow water to the waders. Head for the rocks on either side of the beach, where the *limu* that *honu* like is growing. In no time, you'll see nibbling turtles. Just keep your distance and watch to your heart's content. It can't get easier.

Fortunately for inexperienced snorkelers, the youngest (and cutest) *honu* prefer the shallows. Small turtles like to tuck safely under coral ledges and into holes in the reef. Their shells are fourteen to sixteen inches in length—about large pizza size, although the natural magnification of water makes them seem about one-third bigger. They resemble exquisite gems: burnished brown, highlighted with glistening golds and greens. They also blend in remarkably well with the surrounding corals, so you need a sharp eye.

16

When we first met in 1999, the adorable young *honu* swimming here was one of the smallest turtles we'd ever seen. Akebono, named after the famous sumo wrestler from Hawai'i, returns from getting air and heads back under a ledge, which already shelters another juvenile. You did see the second *honu*, right? They do blend in well.

Once you spot a young turtle, don't dive down to get a better look. Disturbing *honu* is never a good idea, not to mention that it's also illegal. Instead, use patience. Hang around quietly and wait. Eventually, the turtle will want to breathe or will get restless and move. If you haven't upset the little *honu* — something you definitely don't want to do — you might get lucky. An unconcerned turtle will often surface to breathe near enough for a snorkeler to get an excellent look.

A departing turtle usually does so slowly enough for you to tag along at a polite distance. You must not chase the turtle or give the appearance of doing so, because that would be rude and illegal. Besides, if you follow too closely or dive to get nearer, you guarantee that the turtle will just keep swimming. Trust us, *honu* can swim a lot farther and faster than you can. On the other hand, if you show respect for the turtle, you might even get an opportunity to see some foraging. Little turtles sometimes peck at random patches of seaweed as they swim along.

Kayaking with *honu*

Kayaking has become popular in Hawai'i, providing another terrific way to experience *honu*. Almost every guided kayak tour includes an opportunity to see turtles.

Some feature a "Turtle Town" as a stop. This usually means that you'll really be snorkeling to look for *honu*. The kayak is simply the means to get there.

Others take you through areas where your guide knows that you can often see turtles surfacing. While this sort of sighting is definitely interesting, it's also usually brief since you're just passing through.

Another rewarding way to watch *honu* from a kayak is to paddle to an area right above a place where the turtles like to rest. At any given time over such a reef, you are likely to see at least one or two *honu* floating on the surface, occasionally with head raised to take a breath.

The first thing to keep in mind is precisely this point: The turtles need to breathe, and that's why they're on the surface. While you want to get close enough for a good look or perhaps a terrific picture, you also want to stay back far enough to make sure the *honu* are getting a chance to fill their lungs. The obvious conclusion is that you don't paddle in a mad rush straight toward the turtle.

Some turtles are nervous. As soon as they catch sight of your kayak, they'll dive. Others won't dive unless you start to approach. Some turtles

Our kayak allows us to extend the range of turtle observations well beyond the limits that our air supply imposes underwater. We can take the kayak to a resting site that is out of diving range, then don snorkel gear and jump in the water. The *honu* often surface to breathe within camera range, but timing both the raised head and the rise and fall of the waves still makes it tricky to get good pictures.

seem not to care about the kayak at all and will stay on the surface even right next to your boat. Unfortunately, there is no way for you to know in advance which *honu* are the tolerant ones. You therefore need a strategy that gives you the best chance of a close encounter without upsetting any turtles.

First, plan out the area in which you'll do your observing. As always, pay close attention to the current and wind conditions, because they play an important part in your strategy and personal safety. You'll be paddling upcurrent (or upwind) past your observation zone, then drifting back through the area toward the floating *honu*. This lets you keep paddling to a minimum, so there is the least disturbance in the water and therefore little to startle the turtle. Use your paddle only to steer carefully so that you drift past the *honu*, not directly at them. The principle to keep in mind is, "If in doubt, be cautious."

A turtle often stays on the surface a surprisingly long time (especially

When the winds are down and the water is still, we don't have to get in the water to get pictures. The turtles enjoy the calm as much as we do, and so they stay longer at the surface, perhaps soaking up some of that warm Hawaiian sun. This gives us a much better chance to get a profile shot that we can use to identify the *honu*.

if the water is calm), so there is no need to hurry. Furthermore, if you are in the right place, you don't need to dash over to every shell you see. Have some patience, enjoy the ocean, and wait. There'll be another turtle popping up shortly.

Diving with *honu*

Unquestionably, the best way to see *honu* is to use scuba. In Hawai'i, anyone who dives can be certain of seeing a turtle. Because the turtle population has grown so much, the number of places to see turtles increases steadily every year.

Most likely you will be on an escorted dive when you see *honu*. This makes finding the turtles easy. The hard part is making sure that your visit to the *honu* doesn't bother them.

A local dive company takes you and several other divers to the site, usually by boat. The divemaster will explain what you will see and how to behave before you go into the water. Probably one of the first things mentioned will be to show consideration for *honu,* because that way you get longer and closer looks.

Once you're under the water, the divemaster leads you right to an area where turtles congregate. On a dive tour, many *honu* you encounter are likely to be quite familiar with divers. This means that you have an excellent chance to get near without disturbing the *honu* and — if you bring a camera — to take some great pictures. You just need to mind your manners.

If you are with a group of divers, you won't make any friends (human or *honu*) by rushing up to the turtles and annoying them so that they leave. If one turtle flees, others in the vicinity are much more inclined to do the same. The disappearing rear end of a *honu* is not nearly as attractive as the whole turtle resting on the reef.

The first step in showing your respect is to be aware of the turtle's *comfort zone.* Like all animals, *honu* have a comfort zone that varies by individual; get too close and the turtle gets uneasy. If the turtles see you — and usually they will — they'll give a clear indication when your proximity makes them uncomfortable.

Their signals aren't hard to recognize. Watch the head and the flippers. When the head comes up and the turtle turns toward you, treat that as a caution. Stop your approach and wait to see what happens next. The *honu* might just be curious and want a better look at the approaching awkward, bubbly creature. On the other hand, perhaps the turtle is wary but not yet alarmed enough to leave. For caution, you must assume that this is the case and act accordingly: backpedal.

If the flippers start to move, you're probably too close and should back off. If the turtle rises up on its flippers, it's likely already too late — the *honu* is about to depart. Whether the turtle leaves or not, retreat anyway and learn from your mistake.

Honu use two different types of departure: "I'm leaving" and "I'm curious."

In "I'm leaving," the turtle swims directly away, disappearing into the distance. Usually the pace is leisurely, but if you've somehow startled the *honu,* rapid, strong strokes from those huge flippers result in a surprisingly fast flight. This is a bad thing, to be avoided at all costs short of your own safety.

Here is a classic example of a *honu* in alert position. The turtle has risen up from the coral, its neck extended to get a better look, and flippers deployed in preparation for a departure if necessary. (Note that this *honu* is not exercising proper reef etiquette and is "stepping" on the coral.)

In "I'm curious," the *honu* ascends casually to the surface. There, the turtle sips a few breaths of fresh air while studying events below. Pick a spot in sight of the place where the *honu* was resting, settle to the bottom, and wait patiently. Chances are good that the turtle will decide you're harmless and that it's okay to come back. The return path might even include a swoop right over your head, providing a spectacular photo or video opportunity.

If the *honu* are enjoying (or at least tolerating) your company, they stay planted on the bottom, content to watch you watching them. It's really about trust. If you do things right, *honu* trust you so much they'll even turn their backs to you, or go about doing whatever they were doing as if you weren't there.

You'll know you've really got it right if they close their eyes and drift off to sleep in your presence. That's the greatest *honu* compliment of all.

"Sleep?" we hear you ask. Yes, sleep: Many *honu* are primarily nocturnal feeders, so for the ones lying about the reef during the day, you're in their bedroom. Think about that for a minute.

How would you feel if you were just settling down for a comfy sleep and half a dozen or so clumsy, unruly strangers crowded into your room, perhaps even coming up and sitting on the bed next to you? We'll bet you wouldn't be happy.

The remarkable thing about *honu* is that they often *don't* flee at the first sign of divers. This is what makes Hawaiian turtles unique. Almost anywhere else in the world, fright and flight are the turtle's natural responses. Many *honu,* however, have grown to trust humans. There's a good reason for this: we haven't hunted or hurt them since the mid-1970s. It's your duty not to break that trust and to reinforce their confidence by leaving them undisturbed.

Finding *honu* underwater

You and a buddy might also go diving on your own to watch *honu.* Perhaps you already know where to find the turtles, or maybe you're looking for them. If you aren't diving at one of the well-known Turtle Towns, the turtles you encounter are less likely to be used to divers, so probably it will be harder to get close. On the other hand, there will be fewer divers trying to get a look at the turtles, therefore less chance that someone might disturb them — and as always, your goal is untroubled and happy *honu.*

There are a number of things that increase the chances of discovering *honu.* If you've scanned the ocean from shore, you might already have spotted a place where turtles are frequently on the surface for air. If you haven't, once underwater head for a reef, the preferred *honu* haunt. You can often see turtles in other places, but a reef in about thirty to fifty feet of water and one to two hundred yards offshore offers the best odds.

Look for prominent, visually interesting features, such as coral outcroppings, pinnacles, or vertical relief. If *honu* are resting in the area, these are the best places to find them.

Incidentally, this is true away from the reef also. Conspicuous rocks or coral heads or even sunken man-made objects often attract turtles. The

In Hawaiian, *moe* means "to sleep, to lie down," and that's what this turtle does. Moe has slept through every summer since we met him in 1999. We have rarely seen him with open eyelids, and that is no exaggeration! He's slept in almost every position you can imagine, including on his chin. Does this look uncomfortable? Perhaps for you, but obviously not for him.

Big winter waves swept this chair off the beach and a couple of hundred yards from shore, where it was still used for lounging around—but by the *honu*. On nearly every dive we made at this site, the same young turtle was resting in the chair. Underwater man-made objects often become resting habitat for *honu*.

"attractor" doesn't have to be huge, but it will probably be at least as big as the *honu* — maybe a yard long and no less than a couple of feet high.

Although you might find up to half a dozen *honu* around such an object, most of the turtles will be resting on the reef. There you might come across another clue to the presence of turtles: *Turtle Tramples*.

Honu like to settle down into holes on the reef, which they've often made themselves. We call these Turtle Tramples because that's what *honu* do: They trample the coral. A reef with patches of crushed coral is therefore a strong indicator that *honu* have been hanging around. Remember, the reef is the turtle's bedroom, and just like humans, *honu* typically prefer to sleep in the same bed every night — or in this case, day. It doesn't take many sleeps to create a coral crater.

Ledges, caves, and cracks also appeal to *honu*. They like these for the

Hoa, seen at Honokōwai annually since 1997, is a favorite because his calm manner reassures other *honu* that we aren't a threat. If you look right into his eyes, he gazes straight back.

protection they offer. If you find an area with all of the above, you're almost sure to be in the middle of *honu* habitat. You are likely to see turtles.

Once you've found some *honu*, you can start observing their behavior. There's so much to watch for that we gave the topic its own chapter: "The Things *Honu* Do." For now, we'll just leave you with some thoughts about *honu* personalities.

The nature of *honu*

We've learned that *honu* can be almost as varied in temperament as people are. Playful curiosity seems to drive most youngsters, while adults prefer peace and quiet. Females generally are more reserved and cautious than males. Immature males often begin to feel their oats and will pester larger turtles. If there is any physical scuffling, usually one of the combat-

Nani, whose name is Hawaiian for "beautiful, splendid," rests in her favorite place at the Turtle House, framed by a rainbow of bluestriped snappers, known in Hawai'i by their Polynesian name *ta'ape*.

ants is a male. We've concluded that deep inside, however, even the most blustery Hawaiian turtle is really meek at heart.

Yet within these generalizations there are exceptions: young turtles quiet and withdrawn, adults boisterous and inherently obnoxious, and young males with the refined demeanor of an English butler. It's all part of the fascination of observing *honu*.

The eyes are the windows of the soul, it's said. Look into a *honu's* eyes some time. If you've been careful and kind — gained the *honu's* confidence — those eyes will look placidly back at you.

There is no better experience — no happier time — for us than resting with old friends on the ocean bottom.

Heaven, they say, is Up There, with white clouds and halos and angels, but Heaven for us happens underwater, when the late afternoon ocean shimmers electric blue and sunshine ripples gold on the backs of our *honu* friends.

Heaven is being surrounded by sea turtles and, if not being accepted as one of them, at least being accepted. Heaven is where *honu* are.

3 About *Honu*

· ·

It is rather interesting — at least to me — but
then I find everything about turtles interesting.
—Jean Beasley, Topsail Turtle Project,
    Animal Planet Hero of the Year 2007

Why are they called green turtles?

Perhaps the first thing we learned about *honu* is that they are green turtles.
The scientific name for all green turtles is *Chelonia mydas.* Since Hawaiian
green turtles are what scientists call a *population* or *stock,* not a subspecies, they don't have a subspecies name.

   Although their shells often look green, especially when seen underwater, the real reason for their common name is unsettling: Their body fat
is generally green.

   The green turtle is the tastiest of the sea turtles. Coastal-dwelling humans have harvested them for thousands of years. This is why they are
identified so strongly with the color of their body fat.

   Humans have exploited green turtles to the point that they were listed
as either endangered or threatened throughout their range. In Hawai'i
their numbers got dangerously low, but fortunately the state and later the
U.S. government protected the *honu* in time. The once-depleted population has since made a spectacular recovery.

A little sea turtle biology

When we met Clothahump, all we knew about sea turtles was that they
lived in the sea. Clearly, we were in need of a little education. In order to
ask questions about *honu,* we first had to find out how to describe them.
This meant learning the body parts of the turtle.

Thanks to the laws protecting threatened and endangered species, this handsome juvenile is unlikely to become dinner for anyone but a large shark.

We didn't always know the right names, so in the beginning we used terms that seemed suitable to us. Sometimes they were — well — whimsical. Later, even though we had learned the proper terminology, we often used our original names because we like whimsy. Occasionally these terms have slipped into the book. (Don't worry, we don't really think that *honu* have feet.)

Parts of a *honu*

The obvious parts of a *honu* are the head, neck, flippers, shell, and tail. As we learned more about the turtles, certain aspects became more significant to us. For example, the face is the way we can identify individuals, but you won't find the face in any biological description.

THE HEAD. *Honu* heads, in proportion to body size, are smaller than those of other marine turtles. It turns out that a part of that head is an unusually thick skull. It's so thick that there isn't a lot of room for the brain, which in an adult is about the size of a shelled almond. The thickness makes evolutionary sense because *honu* need hard heads. They for-

age for seaweed growing close to or right at the shoreline, where strong waves sometimes heave them headfirst into the rocks.

THE BRAIN. The brain of the green turtle is tiny compared to body size. This leads people to conclude that they act almost entirely on instinct and haven't much intelligence. Scientific evidence, however, indicates otherwise.

There's no doubt that despite its size, the *honu* brain has to do a tremendous amount of processing in order to navigate during nesting migrations. You might think that such a small brain would have little room left for much else, but some scientists aren't so sure.

Dr. Roger Mellgren and Dr. Martha Mann have done experiments showing that green turtles are capable of learning. In one study, the turtles learned that biting on a pipe produced a food reward.

Another test showed that green turtles could learn from each other. First, they taught some young green turtles to find food hidden in a box. Next they introduced turtles that had not been trained. As a control, they

The seaweed in this picture makes the underlying rock look deceptively soft, but the *honu*'s closed eyes and retracted neck tell the real tale: A surge has swept the turtle headfirst into the dinner plate.

placed untrained turtles in a tank with hidden food but with no trained companions.

The untrained turtles with trained companions learned to find the food more quickly than the turtles without trained companions. In other words, turtles can learn by observation.

While not exactly establishing green turtles at the top of the animal IQ chart, these studies do reveal that more than just instinct is at work in that little brain.

THE FACE. *Honu* faces are delightful. Scientists aren't particularly interested in the face as a whole, but we are. One of the first questions that people ask after hearing about our turtle experiences is, "How do you tell them apart?"

The answer is that we've found that every *honu* we've met has a unique face. Many turtles look alike at a casual glance, but this changes if you examine the turtle's "mug shots" closely.

The left and right profiles of the face of a green turtle are made up of several scales. Usually there are around fifteen, but sometimes there are fewer and there can be twenty or more.

You can readily distinguish facial patterns from one another with a little study. Pairing left and right profiles for a turtle lowers the chance that there will be an animal with duplicate markings so much that we consider that likelihood inconsequential.

By the end of 2004, we'd logged more than 750 individual turtles. We're convinced that you can reliably identify individuals using this method. Many *honu* in our database are old acquaintances, seen year after year — all known to us by their distinctive faces.

THE EYES. *Honu* eyes are slate gray, with round, black pupils, giving a tremendous impression of depth. They stand out from the head somewhat, so to us *honu* look like they have big eyebrows.

Scientists tell us that the turtles have excellent underwater vision. Since water is a natural magnifier, this means that above water, *honu* are nearsighted. Scientists have also determined that they are capable of distinguishing some colors, but they are least sensitive to reds. This is why

Facing page: These are some of the left and right profiles that we use in our facial identification database. We encourage you to practice recognizing individual turtles by matching the profiles with faces in our other illustrations.

Honu from A to Z

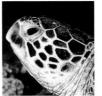

Nui

Akebono

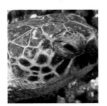

Raphael

Clothahump

Tiamat (122C)

Iakona (H238)

Tutu (U521)

Kimo (U362)

Wana

Makana

Zeus

33

The shimmering of salt flowing from the lachrymal gland is a lot easier to see underwater than it is to photograph. Here the reflection of the shell in the "tears" makes them a little easier to make out.

people who monitor sea turtle nesting at night sometimes use flashlights with red filters.

A popular myth about sea turtles is that the females cry because they will never see their unborn young. It turns out that when sea turtles crawl ashore to nest, they really do weep. The *lachrymal* or salt gland, an interesting feature located behind a sea turtle's eye, removes excess salt from the body. This gland causes "tears" to flow constantly. Underwater, if you look closely you can see this as a rippling in the water around the *honu*'s eye.

When turtles are out of the water, the tears keep the eyes moist. The only time most people see turtles ashore is when they nest. A constant stream of tears flows throughout the long nesting ordeal. These tears tug

The *honu* in the foreground has closed nostrils; the turtle in the background does not. Why do some *honu* open their nostrils underwater? That question still puzzles us.

at the heart and add to the romance of the sea turtle — but they aren't tears of motherly sorrow.

If you go to Hawai'i, you can see *honu* on the beach, happily shedding tears while they enjoy the sun's rays.

THE NOSE. *Honu* noses consist of nares or nasal cavities at the front of the beak, just where a good nose should be. If you get a good look at a turtle nose underwater, you might notice that some of the time these are closed up. You'd expect that, but surprisingly some of the time they are not.

Inside a *honu* nose there is a high concentration of blood vessels. When submerged, a *honu* can (and often does) seal these openings by filling the little veins with blood.

A *honu* is about to scrape up some *limu*. You can clearly see the serrated lower jaw, which should make you grateful that *honu* aren't aggressive.

One of the times when they are not sealed is when the turtle is eating. Before swallowing, the *honu* compresses the seaweed into a little ball by squeezing it against the roof of the mouth. This ejects water and other material through the nostrils — something rather uncomfortable for humans but routine for *honu*, and it is certainly spectacular to see.

We aren't sure why the nostrils aren't sealed the rest of the time underwater. Perhaps they are trying to smell something. Sea turtles have a reputation for a highly developed sense of smell. Research indicates that scent could be an important factor in the famous ability of sea turtles to return to the beach where they were born. Hawaiian turtles have to find a tiny island in the middle of nowhere — four to five hundred miles away from the main Hawaiian Islands where most *honu* live — so anything that helps would be welcome.

THE MOUTH. *Honu* mouths, like those of all sea turtles, don't have teeth. Instead, the lower jaw has a serrated edge, a characteristic unique to green turtles. Their jaws are big and strong, so you certainly wouldn't want to be bitten by one. There's not much chance of that, fortunately, unless you put your hand where it doesn't belong. So why do *honu* need such heavy-duty mouth equipment to eat seaweed?

If you manage to get close to a *honu* foraging in the shallows, you'll see immediately how the serrated jaw helps. You can actually hear the ripping sound as the turtle literally scrapes and shears the *limu* from the rock.

THE EARS. Like other reptiles, *honu* have middle and inner ears but no outer ear or external opening. The tympanum, or eardrum, is thick and is hidden behind a scale on the side of the head.

Humans hear low-frequency sounds underwater by direct transmission through the skull bone. So do sea turtles, but their thick skulls don't transmit sound well. A sound has to be rather loud to have enough energy to get through that heavy bone. Scientists have therefore concluded that sea turtles, including *honu,* hear some low-frequency sounds but have trouble with higher frequencies.

This means that we're free to talk when we're close to a basking *honu* or perhaps watching one from the kayak. The turtles won't hear us. We still must use caution, however. If we create loud or low-frequency sound (by a paddle hitting the kayak, for example, or on land, from rocks clicking together as we walk over them), the *honu* is more likely to pick that up.

THE NECK. We find *honu* necks interesting because a glance at the neck tells us if the turtle is eating well. A healthy, well-nourished *honu* has a thick neck, even when craning for a good look at something.

Honu, like all sea turtles, can't retract their heads completely, although they do pull their heads back at times, such as when they aren't happy. An effect of this inability to retract the head completely is that there is almost no cavity surrounding the neck. This makes sea turtles more streamlined so that they can swim faster and more efficiently.

Some *honu* have an interesting tendency that might be explained by this inability to retract: They often rest with their heads tucked into holes. At first, it looks like the *honu* adheres to the belief usually attributed to the ostrich — namely that if you can't see danger, it can't hurt you. Ostriches don't actually react to danger that way — they tend to run. On the other hand, if you're a *honu,* you really do reduce risk by resting with your head planted deep in a hole. After all, it's the most valuable vulnerable part you've got, and the rest of your vital organs are well armored.

THE SHELL. *Honu* shells, like those of all turtles, are actually the outer part of their skeletons. The inner skeleton is more commonly called the bones.

The top part of the shell is the *carapace.* Comprised of bone, cartilage,

A *honu* swims overhead, providing a terrific view of the plastron. You can also see how the undersides of the flippers and neck are a creamy light color, contributing to the turtle's protective coloration.

and scaly plates, it's the largest and most obvious part of the turtle. For most *honu*, the visible carapace consists of thirteen hard but thin plates called *scutes*, surrounded by a series of smaller scutes called the *marginals.* The entire arrangement takes on a heart or leaflike shape.

A *honu* hatchling has a black carapace. A juvenile has a rich brown carapace with scalloped marginals, streaked with gold, yellow, and tan. As the turtle grows and ages, the scalloping diminishes and the carapace color changes again. Usually it becomes olive, with yellow and brown streaks or speckles.

A covering growth of fine filaments of seaweed can change the appearance of the carapace. You'll often see fish grazing on the algae of a *honu*'s shell, but they never get it completely clean. Except on the newest young turtles on the reef or turtles just back from a breeding migration, some algal growth always remains. It can range from fine green fuzz to a thick brownish mat.

Often you see white or pink patches on a *honu*'s carapace. This is a harmless kind of stony alga called coralline. These patches change slowly enough to provide temporary markings that you can use to recognize individuals, as long as the observations aren't months apart.

The underside or belly of the turtle is the *plastron*. For *honu,* hatchlings have a white plastron that becomes a creamy yellow as a turtle ages, eventually turning a pale orange.

The light color of the plastron, combined with the dark color of the carapace, is part of a natural defense that is common in many animals, a feature called *countershading.* This means that the top and bottom of the animal have different colors.

When light falls on an animal that is countershaded, the overall effect is to help it fade into the background. The plastron, seen from below, blends into the ocean's surface. In fact, not just the plastron but the entire underside of a turtle is lighter in color than the top.

The streaks and speckles of the *honu's* carapace also help the turtle to blend into the surroundings. They create another natural defense called *disruptive coloration.* This means that the outline of the *honu* is visually broken up and harder to see when there is not much light — and even when there is. The sun's rays playing on the shell match the sunshine shimmering on the corals. Coralline patches on the carapace also aid disruptive coloration.

Barnacles help as well. It's not unusual for barnacles to glue themselves firmly to the surface of a turtle. Usually they become attached to the carapace or the plastron, but sometimes they wind up on the head.

The most common (and obvious) barnacle that you will see on *honu* is *Chelonibia testudinaria,* a species that lives exclusively on sea turtle shells. If you spot one on a swimming *honu,* look closely for a feathery stalk. That's the actual barnacle, the little fellow filter-feeding. In general, these hitchhikers don't harm *honu,* but if enough of them collect on a turtle, the extra drag could hinder swimming.

THE FLIPPERS. *Honu* flippers are larger than those in many other populations of green turtles. This is the major physical distinction of the *honu.*

Like all green turtles, *honu* have one claw on each flipper, both front and rear. The claws are impressive, especially those on the front flippers of the males because they are much longer than those of the females. They are also curved inward, which comes in handy when the male is attempting to mate because they make it easier for him to hook his flippers onto the shell of the female.

Although the female's claws are shorter, they are still useful in helping her cling to corals or rocks, especially when feeding in surging shallow water. They also help when she is making her nest, allowing her to excavate into hard ground.

For several years Polzbarney, a juvenile *honu* who was host to several barnacles, lived close to shore at our Honokōwai dive site. Our friend Paul loved snorkeling out to see the young turtle, whom he called Barney because of the barnacles. We already had named another *honu* Barney, so we called this one Polzbarney. Here we were lucky enough to get a picture of one of Polzbarney's barnacle hitchhikers filter-feeding (see inset).

The large front flippers allow *honu* to swim a lot faster than you might think. Most people consider turtles to be slow moving. That's a mistake when it comes to marine turtles, particularly green turtles. When a *honu* wants to move quickly, get out of the way. Their impressive front flippers can take them up to twenty miles per hour for short bursts.

Even more impressive to us are the hind flippers, or as we like to call them, the feet. On a fully mature *honu*, these are *huge!* The feet are surprisingly flexible. You can often see resting *honu* curl them up as though protecting their tails.

This flexibility has uses. Females use the hind flippers as giant scoops when nesting. Males employ them for much the same purpose as their front claws: They wrap them as best they can around the shells of females they are mating and hang on with the claws.

The hind flippers aren't often used for propulsion, but they do play a part in swimming, especially if the turtle is missing a front flipper. Usually, however, the turtles steer with them, using them like oversized rudders. A *honu* can make a quick turn look effortless just by sticking a foot out — striking and fascinating to see.

THE TAIL. *Honu* tails tell a story. The ones on males do, anyway: the story of reaching maturity. When a young *honu* arrives on the reef after the pelagic stage — the time spent far from land in the open ocean after hatching — the tail is tiny, almost invisible under the trailing edge of the carapace. As the *honu* grows, so does the tail, becoming more prominent as the turtle gets older.

When a male *honu* approaches maturity, however, the tail puts on a growth spurt. It gets much thicker and longer, eventually extending well past his hind flippers. This long tail is the only sure-fire external sign of the sex of a turtle.

Females and all immature *honu* have small tails. You can't determine their sex, although a really big turtle with a stubby tail is probably a female.

THE HIND END. To end this description (pardon the pun), we should mention the *cloaca*. All birds, amphibians, and reptiles have one. It's an opening on the underside of the tail that is primarily used to eliminate both solid and liquid waste, but it also plays a prominent role in reproduction.

For the male, it is the opening through which his penis emerges when mating. For the female, it is the orifice that receives sperm from the male and from which she deposits her eggs.

So that's *honu* from back to belly, left to right, and nose to tail.

Honu size

Speaking of nose to tail, how do you measure a sea turtle? If a male's tail can quickly grow to extend past his feet while his shell remains almost unchanged, would it make sense to measure from the tip of the nose to the end of the tail?

Well, no — it wouldn't. For consistency, scientists prefer to measure sea turtles using a large tree caliper to determine the length of the shell along the top center: the spine. This is called the *straight carapace length*. The second choice is to use a flexible tape to measure along the spine, in which case the measurement is called the *curved carapace length.* Curved carapace length is usually about two to six inches longer than straight carapace length, depending on the shape of the carapace and the size of the turtle.

Although green turtles are the largest of the hard-shelled turtles, on average *honu* are a little smaller than their green turtle cousins from other oceans. The shell of the adult *honu* is anywhere from thirty-two to forty inches long, but you can often see individuals with shells as small as fifteen inches. These, of course, are the juveniles. There are smaller *honu,* but you'll only see them at places such as Sea Life Park or Maui Ocean Center because in the wild, the truly tiny turtles are still living their lives far out to sea.

How much do *honu* weigh? On average, an adult weighs about 250 pounds, although they have been known to tip the scales at nearly twice that. A hatchling is just a bit more than one ounce, so it is a marvel that such a little creature becomes the huge *honu* you see out on the reef—all the more incredible because for most of their lives when they live close to shore, the *honu*'s diet is almost exclusively seaweed.

Honu age

In general, larger *honu* are also older *honu*. Unfortunately, there is no consistent relationship between size and age.

Just like humans, *honu* grow faster when they are younger. Unlike humans, after they become adults they continue to grow, just a lot more slowly. Also, turtles lucky enough to settle where there are good pastures to graze tend to grow faster than those in less bountiful areas. Alas, these factors mean that there is no reliable way to determine the exact age of a particular *honu* that wasn't born in captivity.

People frequently ask us, "How long do they live?" This is another mystery. When a creature matures as late as *honu* do and then reproduces only every two to four years, a relatively long natural life seems likely. The latest computer simulations indicate a span of sixty to seventy years—shorter than many people previously thought but still lengthy for a wild animal.

Another common question we're asked is, "How old are they when they start to lay eggs?" Scientists think that some *honu* in the wild start to breed at around twenty to twenty-five years old, depending on factors such as nutrition and water temperature. Estimates suggest that *honu* in marginal areas can reach forty years old or more before they mature.

4 Life as a *Honu*

. . . .

The turtle lives 'twixt plated decks
Which practically conceal its sex.
I think it clever of the turtle
In such a fix to be so fertile.
—Ogden Nash, "The Turtle"

The *honu*'s life cycle

When we met Clothahump, her size impressed us. Her shell was easily fifteen inches long and almost as wide. At the time, that seemed huge, since our previous turtle experiences were limited to freshwater turtles.

Then came a dive that was nearly as memorable as our first encounter with Clothahump. It was overcast, creating a gloom fifty feet underwater. We were exploring new territory, when we began to make out a large mound looming in the distance. As we approached and details became apparent, silhouettes formed: four huge turtles!

For one unforgettable moment, we were transported to prehistoric times, glimpsing giant reptiles: enormous shadows, impressive, regal, and shy. We had found the adults.

Our sense of turtle size suddenly required dramatic adjustment. These turtles were easily eight, maybe ten times as large as Clothahump. We began wondering if Clothahump would get that big. How long would it take? Where did Clothahump come from, anyway?

Eventually we found the answers to most of these questions. The life cycle of the *honu* is similar to that of other green turtles, and scientists have a good grasp of what is involved. We learned a lot about the green turtle's life cycle through old-fashioned research. Books, papers, the Internet, and questioning the experts — these are the tools that must stand in for personal experience when learning about *honu* migration, nesting, and hatching.

Usually *honu* make nests and hatchlings emerge only at the remote French Frigate Shoals, or Mokupāpapa as they are called in Hawaiian. Since the shoals are inside the Northwestern Hawaiian Islands National Monument, where access is strictly controlled, we thought we would always be limited to secondhand knowledge of these events. Then a turtle chose to nest on a beach fifteen minutes away from Honokōwai. Before we get to her remarkable story, however, we need a little background.

Migration and reproduction

Honu, like most sea turtles, spend nearly all of their lives in their coastal foraging grounds. When they get old enough — a minimum of twenty to twenty-five years or so — the urge to breed begins. They leave their home reefs and swim for a month or more to the nesting beach.

Sometime between February and April, many mature *honu* stop foraging in their home pastures. They begin the long swim to the French Frigate Shoals in the Northwestern Hawaiian Islands. Once there, many of them end up going to the sandy islet called East, where they originally hatched. Like all sea turtles, their natal area calls them to come and reproduce, one of a lifelong series of migrations.

Navigation

This presents an interesting puzzle. How do *honu* navigate to that tiny speck in the central North Pacific Ocean, across open water devoid of any kind of markers?

We found two predominant theories concerning the way sea turtles navigate: by scent clues and by detecting the earth's magnetic field. Neither of these ideas rules out the other, and in fact, research indicates that both might play some role.

The idea that they navigate entirely by smell has some weaknesses. Scent trails tend to dissipate and of course couldn't be helpful if the destination was downcurrent, as Mokupāpapa is from the main islands. Also, animals following a scent trail usually change course frequently in order to pick up the smell, but satellite tracking has shown that the turtles maintain a fairly straight course toward their goal. Still, other experiments have shown that when the turtles get close to their target, odors probably do help them reach their destination.

It seems likely that some kind of biological compass helps turtles in long-distance navigation. Like all sea turtles, *honu* have poor vision above water, so scientists suspected that navigating by the stars was almost certainly out of the question. Sure enough, satellite tracks show that they are able to keep a straight course during overcast conditions and moonless nights.

Is it possible that they are sensitive to the earth's magnetic field? Indeed, some scientific experiments have shown that the earth's magnetism probably plays a role in sea turtle navigation; however, that doesn't seem to be the whole explanation.

Honu are probably using a combination of strategies to help them in their journeys, but no one knows which ones or how they are combined. Exactly how they navigate is still one of the great turtle mysteries.

Who goes?

Not every *honu* of breeding age makes the migration. Females typically nest every two or three years, although some might wait four or even longer. *Honu* mothers aren't necessarily regular, so a female might migrate after two years, then three years, then two again. A lot depends on how well momma has been able to feed in the meantime.

Males aren't on a fixed cycle either. Some migrate annually, some every second year. Some go annually for a while, then skip a year. At our dive site, we've noticed several males with this sort of irregular attendance.

A good example is Nui, a young *honu* we first met in 1990. We had the privilege of watching Nui mature into a fine male. He was missing for the first time in the summer of 1996. Then in 1997, he was absent until late in the summer, when he finally showed up looking skinny and tired — just what you'd expect after a long journey. We think he was at the French Frigate Shoals both years. He was home all summer in 1998 but gone again in 1999. We've recorded similar absences ever since, as well as some late summer appearances in which he looked gaunt and exhausted.

We can't say for sure where he goes, but this pattern fits with what is known about *honu* males attending at the shoals.

The tribulations of the male *honu*

When Nui does go to the breeding grounds, he has to compete with other males for the joy and privilege of fertilizing the eggs of one or more of the females. This turns out to be a tremendous challenge.

Once he locates a female that isn't already occupied with another male, he has to mount her. He attempts this by hooking the claws of his front flippers onto the shell of his chosen mate. Then he curls his long tail and hind flippers underneath her to secure his position, hooking his

These white patches are tissue that is healing from the bites of competing males. When Blue, a male we've known since 2000, returned from the French Frigate Shoals in 2005, the edges of his hind flippers looked like this. Notice the missing pieces, bitten away by his competition during courting and mating.

hind claws to the side edges of her carapace. Accomplishing this is hard enough, but a female often becomes unreceptive once she's made her first nest (usually during mid-May). She will attempt to swim away. If that doesn't work, she'll press her hind flippers together tightly and curl up her tail in order to prevent Nui from completing his mission.

To make matters worse, rival males will move in on Nui while he's mounted and attempt to dislodge him. They'll bite at his head, flippers, and tail, often drawing blood and lacerating his skin. As the wounds heal, scabs form that will appear whitish yellow underwater. These provide clues that help you recognize a male newly returned from the nesting grounds.

On males, you'll see most of the wounds at the trailing edges of the flippers, both front and rear. Sometimes females also bear mating scars, but not as many as the males. You might see white scars around the back of a female's neck, where the male sometimes bites while mounted. Such nipping might be a signal to the female to head for the surface. *Honu* mating can last hours, and since the male is holding on with all flippers, she has to bring them both up for air.

Even if successful, Nui might be the daddy of only some of the female's hatchlings. A female might have multiple mates, each fertilizing some of her eggs. Nui has to work hard to perpetuate his genes.

Nui and the other males, having finished their breeding duties long before the ladies have made their final nests, leave the French Frigate Shoals first. The females begin to follow two or more months later. Scientists think that each *honu* migrates alone. The turtles use their little-understood but marvelous navigational abilities to find their way back to the same foraging and resting grounds they started from months before.

Nesting

At the nesting grounds, when a female *honu* is not mating she spends most of her time resting nearby. At intervals ranging from twelve to sixteen days, she crawls ashore. She waits patiently until sunset, then slowly and ponderously drags herself up the beach until she is above the high-water mark. There she begins the grueling task of making a nest — or maybe not.

Sometimes, she turns and crawls back into the ocean. She might be partway up the beach, or she might have actually started to dig. Then, something happens and she abandons the attempt. This is what is called a *false crawl* .

In some parts of the world, such as Florida, a turtle might make a false crawl because a light from shore or some movement, perhaps human, disturbs her. At the French Frigate Shoals, these disruptions are unlikely, but *honu* make false crawls anyway. Perhaps the sand is too dry or wet when she starts to excavate, or the temperature isn't right, or her eggs really aren't ready after all. Only the turtle really knows why.

Because the French Frigate Shoals are remote and require permission to visit, we never expected to be in the close company of a nesting *honu*. Then, in 2000, a special turtle began making nests on one of the main Hawaiian Islands. The turtle had an unusual history, but she was absolutely typical when it came to nesting — except for her choice of a nesting beach.

She first attracted attention because one night in May, she crawled up onto a Maui beach and dug a nest. Mary Jane Grady, who loves *honu* as so many do these days, eventually managed to read and record the number from a metal tag spotted on her flipper. Although she later came to be called names such as Maui Girl, her official name is her tag number: "5690."

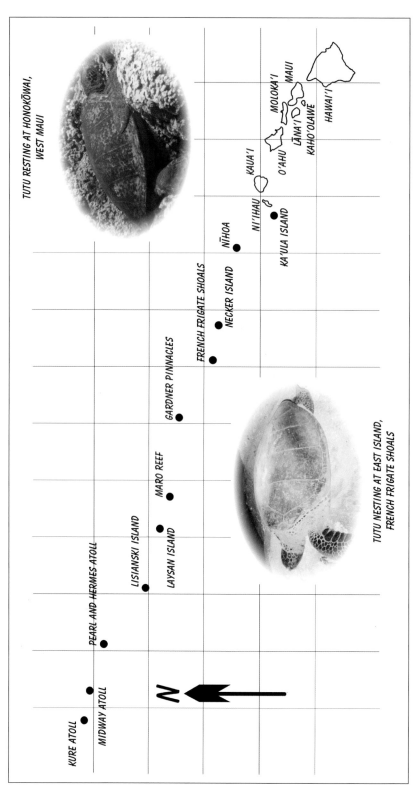

TUTU RESTING AT HONOKŌWAI,
WEST MAUI

TUTU NESTING AT EAST ISLAND,
FRENCH FRIGATE SHOALS

N

KURE ATOLL

MIDWAY ATOLL

PEARL AND HERMES ATOLL

LISIANSKI ISLAND

LAYSAN ISLAND

MARO REEF

GARDNER PINNACLES

FRENCH FRIGATE SHOALS

NECKER ISLAND

NĪHOA

KA'ULA ISLAND

NI'IHAU

KAUA'I

O'AHU

MOLOKA'I

LĀNA'I

MAUI

KAHO'OLAWE

HAWAI'I

The Hawaiian chain extends from the Big Island northwest to Kure Atoll. Most *honu* migrate to the French Frigate Shoals (FFS) to reproduce. Tutu certainly does. Tutu was first recorded nesting at East Island in the FFS in 1988. We first sighted her at Honokōwai, West Maui, in 1990. By 2005, Tutu had migrated and nested at East Island six more times. Every time she returned to her home at Honokōwai. According to George Balazs, Tutu holds the Hawaiian record for the most migrations with both end destinations documented.

She began life in 1980 at the French Frigate Shoals as one of 174 hatchlings that George Balazs brought back as part of a study. Sea Life Park on Oʻahu raised her and the other hatchlings for a year. Then in 1981, George released her off the coast of the Big Island near Hilo. Weighing just seven pounds, she was rather small to be tagged, but George decided to try anyway. He gave her a single tag, which he couldn't attach as well as he'd like. He thought that it probably would be lost, but nothing ventured, nothing gained. He really didn't expect to see her again.

When she reached twenty — just about the youngest a *honu* momma can be — 5690 was ready to breed. As a hatchling she'd never crawled down the beach at East Island, so it wasn't particularly surprising that she didn't head for the French Frigate Shoals with the other *honu* nesters. Her choice of nesting site still surprised everyone, however.

It wasn't a quiet, secluded stretch of sand in some remote part of the Big Island. Rather, it was next to a big hotel in Lahaina, Maui. In the daytime, this is probably Lahaina's busiest beach, and even at night there's plenty of activity. Inevitably, she was spotted. George got a phone call from Mary Jane — and just like that, 5690 was a part of his life again.

By the time 2000 was over, 5690 had made four nests, but we hadn't been lucky enough to see her make any of them. In 2002, however, she returned. This time we were privileged to witness the whole nesting process, with the added thrill of seeing (and helping) George outfit her with a satellite tag. That summer, she made seven nests.

The satellite tracking data revealed that 5690 lived just a few miles northwest of Lahaina, in the Nāpili-Kapalua area, where the *limu* is lush and many other *honu* live — and perhaps significantly, where she was exposed to people, hotels, and lights without any adverse effects.

Returning to Lahaina in 2004, again 5690 made seven nests. She is truly an unusual *honu,* because the average *honu* momma lays only four clutches.

The first time we got the chance to watch 5690, she teased us with a couple of false crawls, but then she decided to get down to business. Like all *honu* mothers, she began by making a *body pit* — a depression about six feet in diameter and about eighteen inches deep. She accomplished this with much determined and vigorous flailing of flippers and slinging of sand, sometimes resembling a mini-sandstorm. Nesting *honu* bulldoze their body pits with power and enthusiasm.

She continued to thrash, her front flippers making loud "thwap" sounds as each stroke finished by slapping against her shell. As she dug

deeper, her movements became more and more precise, until she reached
damp sand and formed the body pit to her liking. Then she switched to
digging purposefully with her feet.

Alternating, 5690 shaped each hind flipper into a huge spoon. Over
and over, she scraped a flipper-cup of sand up from the bottom and put it
in a little pile next to the hole. At the same time, she flicked her opposite
flipper, spraying forward the pile of sand from her previous scoop so that
it couldn't slide back.

She was carving out the *egg chamber* — a flask-shaped hole in the sand
that widened somewhat as it got deeper and had a rounded bottom. Eventually, she began rocking slightly backward each time she scooped so
that she could get the maximum reach from each hind flipper. She was
nearly done. By the time she finished, 5690's hind flippers had dug down
another eighteen inches.

When she felt that the chamber was just right, she began to lay her
eggs. Each egg weighed about an ounce and a half and was a bit larger
than a golf ball. The shells weren't hard like chicken eggs. They were

tough and rubbery, with a small dimple in each, indicating that the egg wasn't quite full so that it would not burst when it dropped. (Eggs that develop normally gradually absorb moisture, and the dimple disappears after a few days.)

Like all nesting sea turtles, once 5690 began to lay — two or three eggs at a time — almost nothing could deter her. Sea turtle researchers refer to this as the "egg-laying trance." At the French Frigate Shoals, nesting monitors take the opportunity to approach the *honu* to check for tags, take measurements, conduct a health examination, and mark her shell with an etching tool and paint, all without harm to the turtle. The *honu* ignores all this, concentrating only on dropping her clutch.

In 5690's case, however, none of this went on. We watched quietly from a respectful distance. We couldn't count her eggs as they dropped, but when researchers excavated 5690's nests in 2002 and 2004, they discovered that she usually laid about 75–90. This is smaller than the typical *honu* nest of 100–110 eggs, but 5690 compensated for the small clutch size by making more nests.

Since she doesn't lay great numbers, 5690 tends to be quick depositing her eggs. From the time when the rocking stopped (indicating the egg chamber was finished) until she resumed movement (signaling that she was covering up), 5690 took only fifteen minutes or so. Other turtles can take up to forty-five minutes.

After all the eggs had been deposited, she began the laborious process of covering the nest. First, 5690 used her feet to push sand into the egg chamber, kneading and packing it carefully. Next the front flippers got going, tossing sand back over the nest. As she did this, 5690 gradually inched forward, obscuring the location of the egg chamber and, of course, the eggs themselves.

Like every nesting sea turtle, concealing the nest was the last motherly act 5690 did for her offspring. It also took the most time. It was an agonizingly slow process that took over two hours and left her drained. Her breathing was labored and audible throughout the ordeal. She took frequent rests, becoming completely immobile until recovering enough energy to continue.

Dawn had long broken before 5690 finished. She took one last extended rest and then began the long crawl back into the ocean, her work done. The buoyancy of the water surely gave her welcome relief. In the way of turtles, her eggs hatched untended, and her offspring had to dig out entirely on their own — something they are well adapted to accomplish.

Above: The egg-laying trance provides an opportunity for identification photos such as this one, taken as 5690 made her fourth nest of 2004. The whitish patch on her back is what remains of the harmless adhesive used to attach the satellite transmitter she carried in 2002. After the batteries ran out of power, George Balazs removed the transmitter during a nesting in 2004, and by 2006 the mark had almost disappeared.

Right: It's not unusual for 5690 to keep us up past dawn, her fourth nest of 2006 being one example. It was 5:37 in the morning before she finished, making this track as she left.

Hatchlings

Because 5690 nests a few minutes away from where we stay on Maui, she has bestowed another *honu* blessing upon us. We have been privileged to see *honu* hatchlings scurry down the beach. Like most things *honu*, this required a lot of patience and persistence on our part.

Once 5690 makes her first nest of the season, we know that twelve to sixteen nights later, she'll make the next one. Hatchlings are not as obliging. Beginning at day fifty-two, we have to check the area around the nest daily because it's difficult to predict when they'll emerge. The primary reason for this is warmth — or lack thereof.

Warm or cool?

Temperature is the earliest influence on a *honu* hatchling. Heat profoundly affects egg development in all cold-blooded animals, including *honu*. Temperature has much to do with how successful 5690's clutch will be: how fast the eggs develop, the amount of activity after hatching, and when the hatchlings emerge from the nest. Eggs developing in warm sand hatch in seven to eight weeks, while those in cool sand take longer.

There's even more riding on temperature: the very sex of a *honu*. Whether the embryo develops into a male or female depends on the sand temperature that prevails during incubation of the eggs. At a point called the *pivotal temperature,* an equal number of males and females are produced. Eggs turn out to be mostly females when they develop in sand warmer than the pivotal temperature. Cooler temperatures result in clutches that are mostly males.

Scientists haven't established the pivotal temperature for *honu* yet, but if we assume that it is similar to that of other green turtles, it's around 84 degrees Fahrenheit.

A lot can therefore depend on whether 5690 makes her nest out in the open or near shade, how deep she digs her nest, or what the weather is like that season. Taken further, a breeding summer of unusually warm, sunny days or one of cool, cloudy days could produce an entire season of hatchlings largely favoring one sex.

The developing *honu*

Whether male or female, a quiet transformation occurs inside each successful egg. A *honu* is happening. Heart, eyes, shell, and flippers form, as do the facial scales and markings that make each creature unique. Yet for all the individuality of the incubation process, to break free from the sand demands the persistent efforts of all.

The temperature of the nest isn't entirely uniform. Eggs around the edges of the nest are influenced more by external environmental conditions than the eggs at the center, which are warmed by the natural heat of their incubating embryotic neighbors. This not only affects the male-female ratio, it also means that some eggs hatch sooner than others.

These "early birds" don't just crawl up through the sand to the surface and make a run for it. If you've ever allowed yourself to be covered in sand, you know that it often isn't possible to free yourself. Sand is deceptively heavy, and sand that has covered an undisturbed egg chamber for two months is not just heavy. In completing her nest, 5690 has packed it down to some degree. Water from rain or waves can compact it further.

At East Island, the relatively small nesting area for *honu* results in a reasonable chance that other turtles have crawled over the nest, compressing the sand even more. But 5690's hatchlings face a different problem: They're buried under a busy beach that thousands and thousands of feet trample during their incubation. Fortunately, when 5690 makes a nest someone ropes off the area almost immediately. The guardian is often a state or federal official, but it's just as likely to be a regular at the beach, such as a surfing instructor or a local resident.

Her hatchlings need all the help they can get, because getting to the surface is no small feat.

Group action

The struggle to the top isn't just about fighting your way *up*. It's also about sifting the sand *down* from above to raise the floor. Creating this sand "elevator" begins as a barely perceptible process, a few grains at a time.

Hatching itself — simply breaking out of the egg — loosens some of the sand in the egg chamber. The eggshell collapses. Since the volume of the unhatched egg is greater than that of the hatchling and the broken shell, this makes room for sand to trickle down. The first little turtles to hatch wriggle about and loosen the sand even more.

Like children forced to share the same small bed, the movement of one youngster stimulates the others. Many hatchlings, squirming around together, move more and more sand. The top and sides of the nest collapse. The sand sifts down to create a new floor. Together, 5690's clutch slowly rises toward freedom.

All of this work doesn't happen quickly. The little turtles take several days to struggle up and out of their nest. The process of raising the bottom of the nest is slow and takes place in fits and starts, with long rests in between.

This group effort maximizes the number of hatchlings that will eventually make the mad dash down the beach. Like all sea turtle hatchlings, *honu* use mass emergence as a survival strategy.

At the French Frigate Shoals, the predators know when the turtle nests begin to hatch, and they're waiting. Being part of a crowd reduces any single individual hatchling's chances of being eaten. Though 5690's hatchlings don't have to face a gauntlet of birds and crabs, they don't know that. Instinct prevails, and they still act together.

When the top turtles are within an inch or so of the surface, temperature comes into play one last time. If the top layer of sand is hot, it means that it is likely daytime and the sun is beating down. The hatchlings stop moving, waiting for the sand to lose its heat. Cool sand means that it is probably night. The hatchlings instinctively know that night is the best time to emerge. After dark, they avoid the potentially deadly heat of the sun, and it's harder for predators to see them.

The great escape

This point in the process gives nest watchers like us the first clue that hatchlings are almost ready to emerge. At the surface, the sand collapses into a shallow funnel shape. Sometimes we can see little *honu* beaks or perhaps the tip of a flipper protruding from the sand.

The first time we saw parts of hatchlings poking out, we got excited. We thought that at any second, 5690's hatchlings would wriggle out and run for the water. We waited. We waited some more. We waited a long, long time — over three hours. The tiny turtles remained completely stationary, with eyes closed. By then, we were wondering whether they were really alive.

They were. This turns out to be a normal behavior, frustrating though it might be to human observers. They were waiting for some signal — just what, we don't know — before the mad dash began. Perhaps they were gathering energy, or perhaps this suspension is necessary for imprinting. Who knows — maybe they just wanted to snooze together for one last time.

Imprinting is the term describing the instant and irrevocable conditioning that happens at a particular stage of life, most often shortly after

Composite of several digital photos with hand-drawn fine detail enhancements. This image represents the experience of 5690's hatchlings on the night they leave the nest. The constellation known to Hawaiians as "Maui's Fishhook" (the tail of Scorpius) greets them from above.

birth. The most widely known example is the way certain bird types bond with an object soon after hatching. This is usually a parent, but it can be an inanimate object or even a human.

When *honu* break out of their eggshells, there is no mother to care for them, but scientists believe that the tiny hatchlings undergo an imprinting process every bit as powerful and permanent as those of baby birds to their mother. Instead of connecting with a caretaker figure, however, sea turtles imprint on a location: their natal beach. In the 1990s, DNA studies finally confirmed the widely held belief that female sea turtles returned to nest on the beach where they hatched. For more than 90 percent of *honu*, this means the islets of Mokupāpapa, and most likely East Island.

Yet almost all *honu* hatchlings, once they erupt from the nest and rush into the ocean, never see or feel East Island until they make their first breeding migration decades later. The intriguing question, therefore, is how do they recognize their natal beach?

Dr. Archie Carr, the father of sea turtle research, hypothesized that hatchlings imprint the smells of the nesting beach during their crawl from nest to ocean. Other scientists have speculated that they might imprint other chemical cues, such as tastes, from the beach or the ocean or both. For this reason, when turtle nests are excavated and hatchlings rescued, they are set down and encouraged to scramble to the water on their own.

Studies of sea turtle navigation suggest another possible form of imprinting: The developing brain could be stamped with a sense of location based on the earth's magnetic field.

No one knows for sure how imprinting works in sea turtles, but it is clear that something tells them where they started life. Imprinting doesn't mean that sea turtles *must* nest on their natal beaches, however. If a nesting beach is destroyed or made unusable, or if imprinting is imperfect, the turtles do nest somewhere else.

Whether 5690's hatchlings were imprinting or not, one instant they were all motionless and the next they were all astir. First the two or three at the surface awoke. Then more and more black bodies struggled free from the sand. For a few moments the horde milled about, then as if by signal, off they charged.

Little *honu* dashed toward the waves. The frantic motions of their tiny flippers revealed an instinctive urgency, compelling them to get off the beach and away from shallow coastal waters quickly, for that's where they're at most risk from predators.

A hatchling, one of 5690's offspring, reaches the ocean. This one was a straggler, rescued during excavation from certain death at the bottom of the nest. If this is a female and her good luck holds, she will return to this beach sometime around 2020 to make her own nests.

No one is certain how newly hatched turtles know which way the ocean lies, but experiments have shown that they seem to take their cue from the brightest and lowest horizon. The ocean surface reflects more light than land, and beaches normally slope toward the water.

This intuition can be the death of a hatchling, because visual clues (bright light) tend to be more important than slope clues. In some parts of the world, this means that they head inland toward buildings and streets. Most *honu,* however, are lucky. There are no lighted structures or streetlights to confuse them in the remote, isolated French Frigate Shoals.

Because she makes nests on a beach with numerous lights in sight, 5690 is a special case. Her hatchlings are at greater risk, but they have offsetting luck. Her nests are known, marked, and monitored. Her offspring have a good chance of getting help should they head in the wrong direction.

There was no need to redirect the little ones we saw emerging. Two or three of them headed off at a bit of an angle, but all of them knew which way to go. We watched in wonder as the tiny turtles scurried through the surf and into the ocean. Within minutes, the surf swept up the last miniature *honu,* and then there was just an occasional black blob on the surface as one of them popped up for air. For us, it was over.

For 5690's hatchlings, it had just begun. Waves close to shore move directly toward the beach, so they instinctively swam straight into the swell. This took them out to sea. Her hatchlings — each unique one — had left the sand for the uncertain life of the open ocean.

The "Lost Years"

Once in the water, their little flippers worked madly in a "swimming frenzy." This is a period of continuous swimming that lasts from twenty-four to forty-eight hours. It's nature's way of getting them far away from the birds, crabs, and reef fish that prey on them and out into the *honu's* nursery: the currents and gyres of the North Pacific.

Where were they headed? Aside from the answer, "Out to sea," nobody really knows. Once they've left the beach upon which they hatched, *honu* begin the oceanic stage of their lives: their "Lost Years."

Dr. Archie Carr coined the term "the Lost Year" to describe the pelagic stage — the mysterious period between the time hatchlings scamper down the beach and when they show up later (and much bigger) at their foraging grounds.

At first, it was the Lost Year because scientists thought that was how long the pelagic stage lasted, but we know today that it lasts multiple years — how many, no one can say. It could vary from species to species or even from turtle to turtle.

What happens during the Lost Years? All we know for sure is that they "go and grow." At this stage in their lives, *honu* are opportunistic feeders. Obviously, at sea they can't get the bottom-growing seaweeds they'll eat later in life. Hatchlings are therefore omnivores, eating macroplankton and anything else they can find in the open ocean. Whatever they eat, they find enough of it to grow from the tiny hatchlings we see disappearing into the ocean to the dinner plate–sized juveniles that show up in nearshore waters.

For *honu,* the latest studies indicate that this period probably lasts from four to six years. Exactly how long, where, and how far they travel remain elusive mysteries. Until someone invents a transmitter tiny enough to attach to a hatchling, this part of the *honu's* life is completely "lost."

"Lost Years" is an excellent term when you think about it. These tiny turtles are lost to scientists and conservationists in the vastness and vagaries of the ocean. Out there, scientists can't study them and conservationists can't protect them. It is easier to find a specific grain of sand on Waikīkī Beach than a palm-sized hatchling in the central North Pacific. Baby *honu* might as well be on the moon! Fortunately, Mother Nature has a way of looking after her own.

New arrivals

Scientists refer to the smallest and youngest sea turtles that you will see on the reef as "recruits." The term refers to the fact that these youngsters are recent arrivals that are just beginning to settle into their new homes, where they can finally touch the bottom. They're "recruited" into the local coastal population from the open ocean.

It probably won't take long for a little *honu* that has decided to stay in a particular underwater area to come to the attention of humans. People who don't venture into the water at all might see a tiny head popping up in the shallows as the youngster comes up to breathe. Snorkelers and divers might spot the little *honu* tucked under a ledge or actively swimming about.

You'll find it easy to recognize recruits. The shell is roughly the diameter of a large pizza — about fourteen inches — but more heart shaped than round. The scutes along the edges are scalloped rather than smooth. The plastron is still pure white. Their shells and skin are remarkably free of the green fuzzy algae and barnacles that eventually collect on the carapace, plastron, and even the head of older, larger *honu*. Put another way, the shell is like a precious stone, with a sunburst pattern of brilliant gold and yellow highlights streaked through a lustrous, warm brown.

As beautiful as recruits are on the outside, they're every bit a gem on the inside too. That's because they have brains that can learn — tiny brains to be sure, but brains that are also curious and can benefit from experience. *Honu* recruits are all promise, so if you encounter one, remember that they are impressionable and still learning. How you behave might have a significant effect on the little *honu*'s future attitude toward humans.

Recruits are surprisingly varied in temperament. When they meet up with a human, the response can range from confident approach to all-out flight, although most youngsters are caught up in a delicate dance between curiosity and caution.

When you think of recruits, think adaptation. This is the time in a *honu*'s life when dramatic changes occur. Their habitat changes from pelagic (the open ocean) to benthic (the ocean bottom). For *honu*, this means waters close to shore. Their diet changes from omnivorous (anything edible) to herbivorous (seaweed, or as *honu* would call it, *limu*). The threats change, too.

Out on the high seas, while they are tiny, the threats are almost entirely of two kinds: large fish that might eat them and, much less likely, garbage that might seem edible but actually can kill them.

Inshore, there are still sharks that can devour them, but recruits have grown enough to be safe from the other big fish they might encounter, such as jacks or barracuda. Now humans or human activities become threats: fishing lines and nets, boats, and of course, pollution from agricultural runoff and sewage.

To humans, recruits are just about the cutest things in the ocean, so it is hard to resist the urge to touch. The turtles can easily avoid contact in most cases, however. The danger to the little *honu* might be minor, but touching and grabbing does teach turtles that humans are not to be trusted and are best avoided, so please resist the urge: Don't touch.

If you are lucky enough to see a recruit, keep your distance and watch. The little *honu* might approach you if you're quiet and nonthreatening. If you're really lucky, you might get to see the youngster at play.

Like kiddies everywhere, young *honu* engage in what we can only interpret as play. We've been delighted to observe this behavior on a number of occasions.

In 1998, we met a little recruit with a shell so spectacular and beautiful that we named the youngster Makana, which is Hawaiian for "gift."

In 1999, we started spotting the tiniest recruit that we'd ever seen — so small that we couldn't help bestowing the name Akebono, after the huge Hawaiian-born sumo wrestler.

One afternoon we watched Makana purposely swim up to a resting Akebono and nudge the smaller *honu* into what certainly looked like a game of tag-and-chase. For several minutes, they circled round each other, occasionally making contact but mostly just chasing each other's tails. They certainly appeared to be enjoying themselves.

We've also seen young *honu* rise up from resting on the bottom and face into a strong current, then swim at just the right speed to hover: *honu* surfing! Why? Perhaps the turtle enjoys it and gets some exercise without having to go anywhere.

Juvenile turtles seem to enjoy play. This is one of our most delightful and memorable observations. In 1999, Honokōwai was blessed with two young turtles, Makana (left) and Akebono (right). It was terrific to watch them play — and funny too. The game never involved chasing each other over large areas of reef but rather a compact circular chase, a dance of round-and-round where one would try to nip the tail of the other. If the chasing turtle got close, the youngster being chased would squirt out of the way at the last second, often tilting the shell. It eventually occurred to us that this game had a serious side when we saw Makana use the same "ballet" on a blacktip reef shark following Makana round-and-round. Watching Makana draw ever-tighter circles, we realized that even *honu* fun could have a practical application.

In some ways, the recruit stage resembles a children's kindergarten. There are lots of lessons but with time off to have a little fun. "Recruit" isn't a precisely defined term, so it isn't easy to say how long this period lasts — but it's short, especially in terms of the potential lifespan of a *honu*. Once the gloss is off the shell, the algae is on the skin, the plastron begins to yellow, and the diet has become predominantly *limu*, the little turtle graduates from kindergarten and joins the big kids in the first grade: the juveniles.

Younger and older *honu*

Turtle biologists in Hawai'i divide *honu* into life stages using the terms "juvenile," "subadult," and "adult," yet they rarely know how old they are. Instead they apply the rough principle that older is bigger. Scientists call *honu* "juveniles" until their straight carapace length reaches sixty-five centimeters, or roughly twenty-five inches. At that stage, *honu* are known as "subadults" until they grow to minimum breeding size, at which point they are called "adults." For *honu,* that's greater than eighty-two centimeters, or just over thirty-two inches.

As turtle watchers, we don't need to be that precise about the divisions. Just think of them as youngsters and oldsters. In fact, the only reason we make the distinction at all is to point out a few differences in behavior.

The young ones are more active. They are more likely to forage during the day, and they seek food more often. Because younger turtles are still developing lung capacity and efficiency, they need to stay closer to the surface and breathe more frequently. As they get older, their lungs develop and the time they can stay beneath the surface gradually increases.

Active turtles need to breathe more often, just like active humans. For example, during feeding *honu* pop their heads up for a quick breath every five or ten minutes. When they are resting on the reef, however, they usually stay down for thirty to forty-five minutes, and sometimes longer. A resting adult *honu* would have little trouble staying under for a couple of hours.

This might explain why younger *honu* usually settle fairly close to shore in a *puka* (Hawaiian for "hole") or under a ledge. The water is shallow, so it's easier to go up to breathe, and they're close to food. Besides, inshore they get more protection from sharks, which tend to remain in deeper water.

Some places are called "developmental habitats" because they attract almost exclusively small, young turtles. The characteristics are ideal for small *honu,* but as the turtles get bigger they outgrow the area. They begin to find that water is too shallow, and they can no longer fit in a *puka.* As *honu* youngsters gradually become oldsters, they start spending the day in deeper water, lying around on the corals most of the time. That's why we call the deeper reefs "resting habitat."

The bulk of their daily activity consists of occasional trips to the surface for air. In our dive area, bigger *honu* don't forage much during the day, usually eating in the late afternoon and early evening or around dawn.

Facing page: We see males attempting to mount other *honu* occasionally, but the turtle on the bottom is almost never female. This was one of the rare exceptions. It wasn't a successful mating, since the incident lasted only a couple of minutes before the pair went to the surface and separated. If the male had achieved his goal, the couple would have stayed together for hours.

Of course, the biggest distinction between youngsters and oldsters is the one that determines the true adults: Mature *honu* migrate to breed.

There is one more behavior that older turtles exhibit. You just might see mounting—a male grabbing another turtle from behind and holding on.

Perhaps you noticed that we haven't mentioned the sex of the mounted turtle. This is because males—at least the ones we've observed—don't discriminate. The *honu* on the bottom *could* be a mature female, but in our experience that probably isn't the case, since most of the mounted pairs we've seen turned out to be males mounting other males. Fortunately for the victim, these episodes are invariably short.

There are also stories told about male turtles attempting to mount divers. We didn't take these seriously until it happened to us. The male turtle approached from behind and tried to grab hold. Why do they do this? We don't know. Perhaps it is a dominance behavior.

There is another interesting aspect to the mounting that we've seen: Other males in the vicinity rush to the scene of the action. How they know something's up, we can't say, but the dash to get there is one of the times when we've seen a *honu* at full speed—and that's impressive. It's much like the way kids rush into the school cafeteria after someone shouts, "Food fight!"

Aside from these few points, you won't notice much difference between the young and old. Dividing them up is useful when scientists want to study them, but for turtle watching it's not necessary. On the other hand, it is useful information if you're reporting a sighting.

Sorting out the sexes

Without the minor surgical procedure of a fiberoptic internal examination and the drawing of blood for analysis, you can't tell the sex of a young *honu*. You have a better chance as they grow older because of the adult male's tail, but it's still hard to tell females from the larger adolescent males. There are hints, but we've been fooled time and again, so we use them with caution.

Male-male mounting turns out to be so common where we dive that when we see two posi-tioned turtles, we assume it's a male-male mount rather than male-female mating. In this picture, the long tails identify all three *honu* as males. The *honu* doing the mounting hangs on with his flippers and claws. The mounted turtle isn't happy and is rolling to dislodge his "friend." Typical of these incidents, a third male has arrived and is doing his best to intrude.

One clue is that females, top to bottom, are thicker than males. Most mature females have a carapace length of at least thirty-five inches, so that's a second indication. If a *honu* is clearly bigger than that *and* looks a bit rotund *and* still has a small tail, we cautiously assume we are dealing with a female.

Another clue is a small tail on a turtle that has been absent for a summer — the sign of a possible nesting migration. Even then, we can't be sure she's female unless we know that the *honu* has been confirmed nesting.

Until 1997, this meant that she would return from the nesting grounds sporting shiny new tags on her flippers. Because *honu* are generally tol-erant, we'd eventually have an opportunity to read the tags without dis-

These pictures show the identifying number 278 that was harmlessly etched on Tutu's shell when she nested in 2004. Engravings like this make the *honu* readily identifiable from a distance both above and below water.

turbance, report the numbers to George Balazs, and get a confirmation that she'd nested.

Then the monitoring team at the French Frigate Shoals abandoned external tags and switched to small internal passive integrated transponder (PIT) tags, which require a special reader. These tags are actually microchips, safely inserted with little discomfort just under the skin of each hind flipper. While they are great for scientists, this means they no longer attach the visible tags that tell us the turtle is probably a female.

The field team also adds an additional marking that helps us, but it isn't permanent. They use a tool to make a small shallow "etched" number about an inch and a half high on the nester's shell, which is then painted white. This is harmless to the turtle, somewhat like clipping your fingernails. For about a year afterward, this marking is visible on her carapace. If you see a *honu* with fresh white letters or numbers on her shell, it is most likely that she nested that summer.

Unfortunately for turtle watchers, eventually the markings fade from view. How quickly this happens depends on several factors, including how much and what type of algae collects on the shell and, of course, since they are like fingernails, the growth rate of the scutes.

5 *Honu* at Home

So happy together
How is the weather
So happy together
We're happy together
— The Turtles, "Happy Together"

Honu homebodies

If *honu* could speak, there'd be no sweeter word than "home."

As soon as we learned to identify individuals, we recognized that *honu* tend to be faithful to the same places. Some *honu* occupy exactly the same spot year after year. Some have two or three favorite places, while others are content just to stay in the same small area.

Our primary dive site stretches about half a mile, north to south. We've found that some of the turtles we know wander up and down within that area, while others stick to one specific reef. We've dived over the same reefs since 1988, and we've known certain *honu* for almost all of that time. For example, Nui and Tutu have been there since 1990.

Honu really do have a sense of place, of home — of their *kuleana*.

The *kuleana* and its *'ohana*

"*Kuleana*" is a Hawaiian word that means, among other things, "small piece of property." When we learned that the same *honu* stayed in the same vicinity year after year, we asked George Balazs if there were terms to describe the area they inhabited. He told us that there was no formal terminology that applied specifically to marine turtles. He then put forth an excellent suggestion: *kuleana*. When we heard the word, we immediately felt that his choice was exactly right.

We also needed a word to describe the group of *honu* inhabiting a

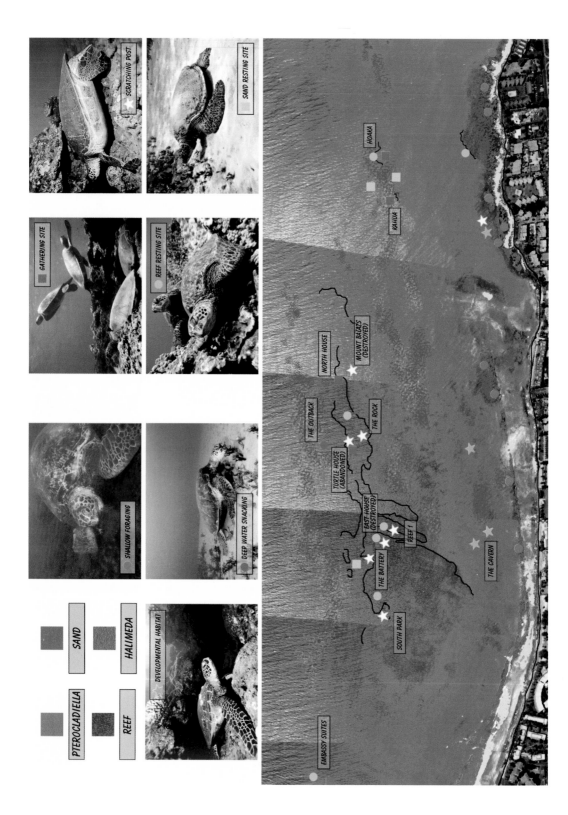

SCRATCHING POST ★

SAND RESTING SITE ■

GATHERING SITE ■

REEF RESTING SITE ●

SHALLOW FORAGING ●

DEEP WATER SNACKING ●

DEVELOPMENTAL HABITAT

PTEROCLADIELLA

SAND

REEF

HALIMEDA

EMBASSY SUITES

HOAKA

KAHUA

NORTH HOUSE

THE OUTBACK

MOUNT BAZZS (DESTROYED)

THE ROCK

TURTLE HOUSE (ABANDONED)

EAST HOUSE (DESTROYED)

REEF 1

THE BATTERY

SOUTH PARK

THE CAVERN

kuleana. As before, George told us that there was no formal term, but again he had an excellent suggestion: *'ohana.* This is a Hawaiian word that is often used to mean "family" but also to describe groups that share an experience, such as a high school class or a softball team. While *honu* residing in the same territory might indeed be related — the Hawaiian breeding population is still relatively small — they aren't really a family, but they *are* an *'ohana.*

A prime *kuleana* attracts and keeps a large *'ohana.* It is divided into two distinct parts: the resting area and the foraging area.

The resting area

The resting area contains one or more extensive reefs on which *honu* can rest and peer out at their world. It's these reefs that contain the clues that tell you that you've come across a *kuleana:* Turtle Tramples.

The *honu* are hard on their *kuleana.* Corals are fragile and turtles are not. Certain types of coral — notably the finger coral known as *Porites compressa* — break easily under the weight of a turtle. Since the *honu* typically prefer to rest in the same places day after day, they often wind up making holes littered at the bottom with crushed coral, which we have dubbed Turtle Tramples because that's what turtles do.

The effect of turtles taking off and landing atop coral leaves little to the imagination, especially when you consider that many *honu* weigh over two hundred pounds. The *honu* can do damage even when resting. Just lying around breaks and grinds brittle finger corals into rubble.

Lobe corals *(Porites lobata)* and rice corals *(Montipora capitata)* are gradually worn away. The *honu* like to lie in the hollows and depressions of this kind of coral, and the constant abrasion eventually rubs the coral surface smooth.

All this wear and tear is the consequence of the preference the *honu* have for lying in "potholes," as the Hawaiian fishermen and divers sometimes refer to them. Not all potholes are formed by *honu,* but some are. Regardless of how a pothole came to be, however, the turtles like to use it.

Wana settles back into in her favorite Turtle Trample after surfacing for air. When this picture was taken, she had been using this same place for three summers, but she didn't make this Trample. Other *honu* rested here long before Wana arrived at Honokōwai, and some of them still share it with her.

The foraging area

A good *kuleana*'s foraging area provides plenty of *limu* and is close to the resting site. Just like us, *honu* prefer the convenience of dinners close by.

The foraging area is most likely to be along a stretch of rocky coastline that hosts the red *limu loloa (Pterocladiella)* that is one of their preferred foods.

The *honu* eat several other kinds of *limu* as well. Here's a *honu* dinner menu, in no particular order: *Ulva fasciata* or *pālahalaha*, *Codium edule* or *wāwaeʻiole*, and *C. reediae* or *ʻaalaʻula*. Seaweeds without Hawaiian names include *Amansia glomerata*, *Spyridia filamentosa*, *Acanthophora spicifera*, and *Hypnea musciformis*. (The last two of these are alien; that

Here's Tamu scratching his plastron in 1996, when he still had an immature and short tail. He's using a chunk of coral that is abrasive enough to satisfy his needs. The bad news is that constant use of the same piece eventually kills the coral polyps. Hours and hours of *honu* rubbing grind the coral into rubble. Once it is dead, algae begin growing over the debris, making it useless for scratching purposes. Eventually the *honu* abandon the site completely.

is, they are not native to Hawai'i.) All of these species grow in shallow water, usually close to shore.

Scratching posts

Aside from good foraging and resting areas, a *kuleana* should have another prized *honu* amenity: large coral heads or volcanic rock ledges on which, beside which, and under which *honu* can rub or scratch.

One of the more intriguing things you'll likely see a *honu* do is scratch. They scratch the tops of their shells, the bottoms of their shells, their soft body parts, their throats, the tops of their heads, and even their rear ends. From this we've concluded that there isn't a part of a sea turtle that they don't scratch sooner or later.

The corals of a *kuleana* bear testimony to the *honu*'s urge to scratch. Years of rubbing will wear portions of coral heads smooth. With their strong front flippers, *honu* will pull themselves forward, dragging their plastrons over the sandpaper-like corals.

Hind feet are surprisingly supple and can grasp things to the rear, allowing the *honu* to reverse gear over the same set of corals. A scratching sea turtle observed from above makes for one remarkable sight: front flippers raking forward, rear flippers returning the *honu* to its original position, accompanied by side-to-side shimmying. It looks for all the world like an old jalopy negotiating a washboard country road as seen from a helicopter.

That's just what they do for their bellies; the backs of the *honu* get scratched, too. *Honu* like to find a rocky overhang that is just the right height to crawl under and rub the top of the shell. These scratching posts are often temporary, however. *Honu* are powerful animals. The forces they generate by rising up eventually can be too much for the outcropping. If so, it cracks, then collapses, and another *honu* scratching post is gone.

One question that still puzzles scientists is *why* they scratch. Humans and many other animals scratch because they itch, but this seems unlikely for a carapace or plastron. (After all, your fingernails don't itch, do they?) Might scratching be a part of cleaning? Nobody knows.

Cleaning stations

A *kuleana* also is likely to have one or more "cleaning stations" — places where turtles cluster while various kinds of reef fish groom them.

A *honu* snuggled away in a Turtle Trample often becomes host to a reef party. A variety of fish gather round because to them, a turtle is the reliable host for a meal: algae on the shell, for example, and even some

Facing page: When the top picture was taken in 1998, Tamu's tail had revealed him to be a male. He still liked to scratch, and here he satisfies a major itch under the lip of a large coral head at East House. For a few years, East House was a major *honu* attraction because the coral was at the perfect height to rub and scratch a *honu* carapace. You might find it interesting to match the patterns on Tamu's hind flippers with those in the photo of him scratching in 1996. These patterns are one way we use to identify individual *honu*. The bottom picture shows that by the summer of 1999, East House had been toppled. The massive overhanging coral proved to be no match for the forces created by the *honu* scratching their carapaces. Because it still provided opportunities for turtles to scratch their plastrons, the *honu* continued to be attracted to it.

skin parasites you'd rather not read about. There are two basic types of cleaner fish: herbivorous grazers and parasite eaters.

The herbivores typically swim and graze in schools. They mostly eat algae on the *honu*'s shell, although it's not uncommon for them to browse on the head or flippers as well. Where we dive, the goldring surgeonfish *(Ctenochaetus strigosus,* Hawaiian name *kole)* is the most common algae cleaner. We also see other herbivores such as convict and yellow tangs helping out. These cleaners eat with a sucking/scraping action, but the sensation doesn't seem to bother the turtles. Since the *honu* seem to enjoy scratching their shells so much, we suspect that if they can feel it at all, they probably enjoy the stimulation.

At times, one of the *honu* drifts slowly above the reef in a presentation posture, while the grazers work. The head and all four limbs hang down, completely exposing body parts for cleaning, and occasionally even the eyelids droop shut. Such a turtle looks even more relaxed and blissful than a *honu* sleeping on the bottom.

The parasite eaters are a completely different story. They nip rather than scrape and usually target skin barnacles, small worms, and other parasites found on the leathery skin of the turtles. The *honu* often flinch when one of these cleaners bites, so we suspect that there's a little pain involved. Still, a reduced parasite load is the reward for putting up with the discomfort, much like the way we tolerate getting our teeth cleaned by a dentist.

The saddleback wrasse *(Thalassoma duperrey,* Hawaiian name *hīnālea lau wili)*, the most obvious parasite cleaner, is a colorful and fast-moving fish. It's rather likeable until you see one snap at a turtle's eye.

The other common parasite eater is the whitespotted toby *(Canthigaster jactator,* no Hawaiian name). This tiny fish likes to take cover under the turtle it feeds on. You often won't see it until the turtle, annoyed by the biting, gets up to leave. In the waters off Honokōwai, it's difficult to find a *honu* without a toby skulking underneath.

All of these species and more will gather where turtles like to rest. The resulting cleaning activity explains why the place becomes known as a cleaning station. We used to think that the cleaners were the reason the *honu* gathered in particular locations, but years of observation have changed our opinion. We've seen the turtles cluster around many spots where there aren't cleaners. On the other hand, we've seen the algae grazers gradually gravitate to groups of resting *honu*. We've concluded that the *honu* create their own cleaning stations.

The turtle in the foreground is Iakopa, who was tagged at Kīholo on the South Kohala coast of the Big Island. In 2001 Iakopa was reported at Olowalu on Maui and by 2003 had taken up residence at Honokōwai. Although *honu* are typically strongly attached to their foraging area, as they grow larger some of them do move about, especially those from the west coast of the Big Island. Iakopa is the Hawaiian form for Jacob. Jacob Almanza felt a special connection with the *honu*. He was just twenty-two when cancer took him from his family. May his spirit be with the *honu* forever.

Moving between *kuleana*

We've mentioned developmental habitats. That kind of *kuleana* attracts recruits but can't meet the needs of older, larger turtles, so as the youngsters mature they must eventually move on. Other *kuleana* include both developmental and adult resting habitats, in which case you'll see *honu* of all sizes. Kāne'ohe Bay, O'ahu, and our own dive site, Honokōwai, Maui, are good examples. Many of the *honu* growing up in areas like these will stay in the vicinity after they mature, although others relocate. When they do, the search for a new *kuleana* — probably their home for the rest of their lives — can take them a long way away.

From some of the tags we've read, we know that turtles arrive in West Maui from Moloka'i and the Kona Coast of the island of Hawai'i. Although the numbers aren't high, George Balazs has collected other tag recoveries documenting cross-channel movements from one island to another.

Are these migrants, seeking new places to settle, or nomads, continually wandering? In our experience, once we've identified a new arrival we'll encounter that turtle again and again. It's true that there are constantly roving *honu,* of course, but there's a lot of evidence for strong site fidelity once they've found a place to their ultimate liking.

After they've settled into a *kuleana,* what do the *honu* do? They mostly do the same things we do at home: eat and sleep. They just tend to do it backwards from us.

Honu at rest

During the day, most *honu* seem to enjoy nothing more than sprawling on the ocean bottom. They could just be lazy, but reptiles often follow periods of eating with rest in order to aid digestion, so we prefer to think that they're hard at work processing their meal.

In any case, once a *honu* has settled down, it usually isn't long before the eyes droop and eventually close and the turtle dozes off.

Sleeping *honu* are beautiful to see, especially when they attract a cloud of colorful cleaners. Aside from that, there isn't much to say about an inactive *honu,* except "Please do not disturb!"

Honu feeding

Once you've seen both, you'd probably agree that foraging is more interesting to observe than resting. Most *honu* feed in about three to six feet of water. Occasionally you might see the *honu* grabbing a bite or two in deeper water. In our experience, this is most likely to happen right after a late afternoon visit to the surface for air. They are probably getting hungry as dinnertime grows near, so instead of heading directly back to the reef, turtles sometimes settle on the bottom amidst some algae such as *Amansia glomerata* and proceed to nibble.

We correlate this kind of foraging to cattle grazing. The turtle extends the neck, craning to eat whatever is within reach before crawling along just far enough to find more.

When we saw red oozing from a turtle's mouth and nostrils for the first time, it alarmed us. We inspected the animal closely for any sign of injury, thinking that perhaps the *honu* had swallowed a fishing hook or worse. We now view these red events for what they actually are: *turtle burps!* This portly *honu* is still busy digesting a dinner of *Pterocladiella,* the red seaweed that gives a *honu* red burps.

Lomi (tag U164) has just come back from the surface, but before settling down to rest she decides to have a little snack. Although it looks like she's eating sand, she's actually using her serrated jaw to scrape up some of the red seaweed *Amansia glomerata,* which is often covered in sand and silt. Midafternoon foraging in forty feet of water or more is not particularly unusual, but it seldom lasts more than a few mouthfuls. Most *honu* foraging takes place much closer to shore.

We don't see the *honu* spending a long time at this sort of eating. After a few minutes, the turtle usually heads back to the reef for another nap. That's why we think of it as snacking, like your trip to the fridge when you get up to stretch during a TV commercial.

In the shallows, we liken the *honu* to chickens. They hunt and peck, because the surge of the waves won't allow them to settle down and graze. Instead, feeding is an active and challenging process that involves flippers flailing just to remain right side up as the *honu* try to get at the *limu.*

Honu food processing

We have been lucky enough to be able to do some undisruptive snorkeling with the Honokōwai *honu* around sunset while they forage. We can only admire their tenacity. They are determined and focused feeders. They rip and scrape *loloa* from the rocks, timing their mouthfuls with the ocean's swell. Waves can leave a foraging *honu* momentarily stranded on the rocks or even swept upside down, tail saluting the sky.

Sooner or later, the *honu* snatches a mouthful of seaweed. The turtle then presses the material against the roof of the mouth, ejecting water and loose debris through the nostrils, and swallows.

The *limu* is first stored in a *crop* that holds food for later digestion. It is located along the esophagus before the stomach and is actually an expansion (ballooning) of the esophagus. It's not clear what purpose it serves, but it does provide an interesting twist to our comparison of the *honu* to chickens while feeding. That's because the crop is primarily a feature of avian digestive systems. In fact, of all the species of marine turtles only some Pacific greens seem to have a crop, such as the *honu* and the Australian green turtles. Perhaps most interesting of all is that scientists don't know why this is so.

From the crop the seaweed moves into the stomach, where actual digestion starts, and then on to the intestines, where it is completed. Eventually the waste is formed into fecal pellets and discharged through the cloaca. For us, this is where it gets interesting.

There is a line that separates the casual observer of an animal from the truly involved. You cross it when you become interested in their waste matter, which we'll politely refer to as fecal pellets. We confess that we carry plastic bags on our dives for the sole purpose of collecting *honu* droppings — brownish-green nuggets that hold a gold mine of *honu* information. Fecal pellets are a way to gain at least partial insight into what the *honu* are eating. Thankfully, we're just collectors; others do the analysis.

It's easy to recognize *honu* fecal pellets. Underwater, they look much like the human version, except they're greenish. If you find one lying on the bottom, look around. It's usually only a matter of time until you find the entire turtle.

Incidentally, if you're watching a turtle underwater and you notice bubbles coming from the hindquarters, watch carefully for a bowel movement. We've noticed that the *honu* often pass gas just before more solid matter emerges.

A foraging site is another good place to find *honu* pellets, but only around feeding time. The wave energy close to shore tends to break them up and remove them or sometimes wash them onto the beach. When that happens, *honu* pellets dry out and turn dark brown or blackish. For-

A truly dedicated student of the ways and habits of *honu* displays a treasured find: a large, fresh, intact pellet of completely processed *honu* forage. A seaweed expert can analyze this sample and see what went into the front end of the turtle to result in this specimen emerging from the back end.

tunately, unlike the droppings of most land animals, the *honu* version doesn't smell or attract flies (perhaps a benefit of being vegetarian).

So you don't even have to go into the water to collect *honu* fecal pellets should you be so inclined, which we realize you most likely aren't. Just walk the beach near a foraging site and look down.

Night feeding

Casual observation easily reveals that many *honu* approach shore to feed around sunset, and that they are also feeding at dawn. Do they feed all night? George Balazs conceived a clever study to find out.

First, he used ordinary dog collars and pouches specially sewn by his wife, Linda, to fashion holders for time/depth recorders (TDRs), small electronic devices that track and store the number of minutes a turtle spends at the surface and at a particular depth.

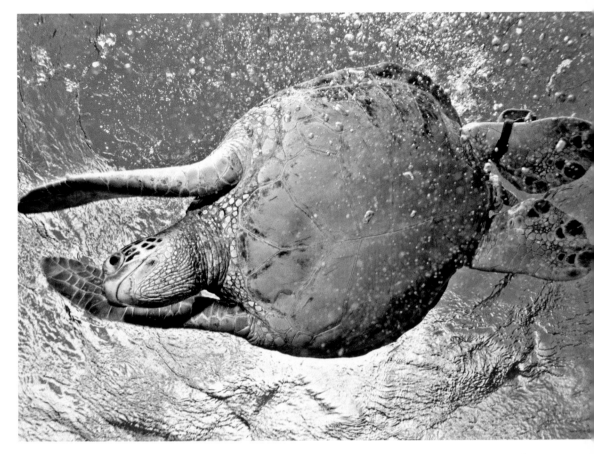

Wana was a test subject for George Balazs' 2002 time/depth recorder (TDR) study. The small electronic device, temporarily attached to her left hind flipper, recorded depth every minute. In this way, we could determine her diving behavior over a twenty-four-hour period. This particular study confirmed that in the evening, Wana (like others in her 'ohana) ventured inshore to feed. Here Wana had just finished some surface basking, filled her lungs with air, and with an impressive burst of powerful flippers headed straight back to her Turtle Trample.

Next, we directed George to certain *honu* at our dive site, thirty feet underwater. These were adults, the size class whose eating habits he knew least about. We selected turtles that we knew he could approach without upset and that we could rely on finding later. George easily attached the TDRs to each chosen *honu*'s hind flipper. The design made it simple for us to unclip and retrieve them when George instructed us to do so. George's colleague Marc Rice, of Hawaii Preparatory Academy, then downloaded the data for analysis. After collecting data on several turtles during two separate summers, we finally got some behavioral insights.

Typical *honu* foraging takes place around dusk and dawn in shallow water. This *honu* is feeding at a place we call the Pantry because the lush beds of *loloa* growing there often attract five or six turtles at a time. Pictures like this are difficult to get because the light is poor at sunset and the water close to shore is often murky due to the surge of the waves and percolation of fresh springwater.

The *honu* we studied spent the day resting for lengthy periods in water around thirty feet or more deep. When they surfaced to breathe, the time spent on top of the water was usually measured in minutes. Of course, we already knew this pattern from diving with the *'ohana* at our dive site.

Right around sunset, when we rarely dive, the TDRs showed that the turtles left the deeper water of the reef. For about half an hour, the *honu* were near or at the surface, the pattern you would see if they were swimming somewhere.

They eventually wound up in water that was typically less than three feet deep. The TDR printouts revealed shortened intervals between breaths, exactly as you would expect when the turtle is more active and therefore using more air. They spent less than a minute at the surface, again consistent with foraging behavior. The *honu* repeated this pattern in the early morning around sunrise.

Soon after sunset, however, some turtles moved into deeper water and went into a resting pattern: twenty to thirty minutes at depth, broken by intervals longer than a minute at the surface. Clearly, some turtles did not feed all night. Some, however, did.

We have more puzzles now than when we started. Where do the turtles that don't feed all night go? The data did not show them swimming back to their known *kuleana* resting area. Besides, the few times that we'd worked up the courage to dive at night, we didn't sight a single *honu* in their daytime haunts. We assume the daytime feeders sleep at night, but where? Do they feed all day, or just part of the time? Do they retreat to the reefs after dark, reversing the pattern we've seen elsewhere? Nobody knows. Perhaps most intriguing, since turtles at some locations in Hawai'i (such as Laniākea on O'ahu) have switched to daytime feeding, why do some Honokōwai *honu* prefer to feed at night? The nocturnal activities of the *honu* are still mostly mysteries.

We don't know if these questions can be answered. We do know George Balazs, however. If there's a way to solve these puzzles, he'll find it.

The faithful *honu*

Over the years, several of the *honu* we've met have become long-term "friends." We've seen them summer after summer, often resting in exactly the same place where we last saw them the previous year.

Mature *honu* go missing some summers because they migrate to the breeding grounds. We worry about them, but they usually manage to find their way back. As we've described, the *honu* have strong site fidelity — they're faithful to their *kuleana*.

Still, some do locate elsewhere, particularly the males. We are more likely to lose track of newly mature males at Honokōwai than the adolescents or females. Although we see many males year after year, some move on. We don't know why the absentees go, but we have resighted a few of these missing males a few miles away both up and down the West Maui coast. Females and adolescents, on the other hand, tend to stay.

Since the *honu* are highly loyal to their breeding and resting sites, you might conclude that they also like to feed in the same places every day. As far as we can tell, you'd be right.

Once we discovered where many of the Honokōwai *honu* were foraging, we quickly determined that the same turtles usually showed up at feeding time. What we found interesting, though, was that not all the *honu* we saw at the resting site came to the foraging site that we were observing.

Conversely, some of the *honu* that we identified at the foraging site were turtles that we hadn't seen resting on the reef. Apparently, different *'ohana* share common foraging pasture.

There's also no doubt that the Honokōwai *kuleana* holds more than one place where turtles feed. Perhaps the *'ohana* spreads out instinctively as a strategy to avoid the sharks that might be attracted to a concentrated group of turtles. Maybe they are avoiding the overgrazing that could occur if they all ate together. It could be as simple as a matter of taste: Some *honu* prefer the *limu* in one location to that of another.

The foraging habits of *honu* still pose many unanswered questions. As scientists are fond of saying, "More research is needed."

Social structure

When we explain the *kuleana* and *'ohana* concepts to people, they want to know whether the *honu* have a social order. Do they form friendships? Do they even recognize each other? Is there any sort of hierarchy? Does one turtle dominate?

We often see the same turtles resting close to one another, but we don't think they do so out of friendship. It is much more likely that this happens because they share the same preference for a particular part of the reef.

They couldn't become friends unless they recognized each other as individuals. We would like to believe that they can do this because we've regarded them as individuals ever since we learned to tell them apart. Aside from wishful thinking, however, we have no evidence that *honu* have the awareness they would need to distinguish one from another.

In fact, the turtles don't even seem to have a pecking order. In other words, we have never seen evidence of one *honu* consistently dominating another, let alone a whole group. Even size is not a reliable predictor of which turtle will prevail in a confrontation — or for that matter, which one will be the aggressor. At times, we've seen the smaller *honu* chase off the larger.

Can't we all just get along?

When we first encountered the *honu,* they tended to cluster around a place that we came to call the Turtle House. We didn't know anything about their social habits, so we thought that what we were seeing was typi-

This is the *mauka* (landward) side of the Turtle House in 1996. While Limu snoozes on a coral head fractured because too many turtles used it as a scratching post, a juvenile rubs against a fallen fragment. It is exactly this sort of activity that eventually destroyed the place. The original Turtle House attracted as many as a dozen turtles at a time. The constant comings and goings, the scratching and rubbing, and even just the resting pummeled these corals to rubble. Just five years after this photo was taken, the area was drastically different. The coral head that dominates this picture is now a mere shell. The surrounding rubble is coated with brown algae, and the ocean floor here is populated primarily by sea urchins. The Turtle House is now an unrecognizable ghost town. Even the fish have moved on. The Turtle House as we knew it exists no more. Such is the dynamic nature of the *honu*'s environment. Fortunately, Hawaiian turtles are highly adaptable, surviving all manner of changes.

cal. There were usually five or six turtles resting together in what seemed to be blissful harmony.

Later we learned that gatherings of *honu* were something new. One reason was that previously, there just weren't that many *honu*. Their numbers didn't begin to increase until the mid-1980s, so they simply weren't as likely to encounter one another and cluster.

The end of turtle hunting in the mid-1970s is another probable factor. Groups of *honu* would have been prime targets for hunters. Once the killing of turtles stopped, it was safe for the *honu* to get together.

This new tendency to congregate inevitably resulted in competition for the prime resting spots. Some *honu* occupy locations that are the envy of others. When one turtle challenges another for a coveted place — and it happens a lot more often than we originally thought — the *honu* get snippy.

Honu can do damage to each other, as shown by their mating scars, but squabbles over spots aren't nearly as serious. Landing atop one another to claim a space is more likely than biting. Even the bites we've seen are really just brief, gentle nibbles, similar to the way you'd tap someone on the shoulder as a nonverbal, "Excuse me, but . . ."

Occasionally we've observed fights that don't appear to be turf tussles. The *honu* go face-to-face, mouths open and snapping. The bouts are brief, and we've never seen any injury result from them. Why they quarrel like this, we can't say.

One thing we've never seen *honu* fight over is food. At the foraging site, *honu* concentrate on eating and struggling with the ocean surge. The turtles will feed close together, sometimes even getting tossed into one another when a wave breaks, all without even the slightest hint of discord.

6 The Things *Honu* Do

. . . .

You don't move around among a
different species for most of your life
without learning to read a lot of their
body language, especially since it's in
such large print.
— Terry Pratchett, *Lords and Ladies*

Honu behavior

The only way to observe most *honu* behavior is to dive with them using
scuba. Even then, you will need many dives and some patience before
patterns begin to emerge.

Since we've been lucky enough to dive for many years with the same
turtles, we've collected pictures and videotape of *honu* engaging in all
sorts of activities. We've studied these images and watched carefully in
the water, and in this chapter we want to describe what we've seen. We'll
tell scuba divers what to look for and give nondivers some idea of what
honu are up to down there.

Honu in motion

Before we describe what we've seen *honu* do, we need to provide a little
background about the language we use to describe *honu* as they swim.

Like most people, we used to think of turtles as slow, ungainly creatures. Then we saw a swimming *honu*. Their similarity to giant, graceful
birds struck us immediately. That's why we often use avian terms in our
descriptions.

We say that a swimming turtle is "in flight." Turtles getting up to leave
are "taking off." Swimming *honu* "bank" to turn and "glide" to a "landing."
Admittedly, landings are the least elegant part of the *honu* repertoire —
they'll bump and scrape bottom, but that's true of some pilots, too.

Later we discovered how natural our inclination was. After all, both air and water are fluids, and the way birds and *honu* move through them is remarkably similar. The turtles use their front flippers to swim in almost the same way that birds use their wings to fly. When you see a *honu* in silhouette from below, the resemblance of the forelimbs to the wings of gliding birds such as frigate birds and terns is obvious.

In fact, scientists commonly use terminology borrowed from aviation to describe swimming marine turtles. Further, the shell of the *honu* has evolved to a streamlined wing shape that reduces drag and increases lift, so the comparison to flying is as appropriate as is it irresistible.

The observations

Assume that you've found a *honu kuleana*. You've mastered the techniques of recognizing when *honu* are uncomfortable and making them at ease. The turtles begin to trust you. What will be your reward?

As soon as the turtles are satisfied that you aren't a threat, they'll go back to their normal behavior. Your reward is to be ignored.

This is exactly what you want, because it lets you observe the daily life of *honu,* which can be interesting, educational, fun, and even funny to watch. If you're careful and really lucky, you'll win the ultimate prize: being treated like another turtle.

While we don't believe for a minute that *honu* mistake us for other sea turtles, we do know that they often treat us the same way that they treat their own kind — or at least like something big that is of no threat or consequence to them.

For example, we've noticed that a turtle swimming over the reef often changes course to get a look at a resting turtle. We therefore find a place to settle where we won't damage the reef, then stay quiet, exercise patience, and just observe. There's a reasonable chance that a turtle will approach.

If the *honu* is just curious, when this happens all you need do is keep your position and enjoy the encounter. Most of the time, the turtle takes a good look, decides you're not really that interesting, and continues onward. There are occasional exceptions, however.

One common "fly-by" involves an approach so close that there is slight contact, such as a gentle brush with a flipper. Perhaps a turtle will try this with you. If so, just relax and let it happen. It will be brief and surprisingly delicate, and you'll have a great story to tell your friends.

Once in a while, you might find that you've occupied the part of the reef that the approaching *honu* wants. Some turtles think they can use the same intimidation tactics on you that they use on other turtles. If you meet a *honu* that is determined to bully you, here's what to remember.

First, the right thing to do is just let the turtle win and give way. It's the turtle's home, not yours. Since you're an uninvited guest, please mind your manners and move well aside.

Next, keep in mind that *honu* don't attack. Instead, they use their size and natural armor plating to get what they're after. They might push you or settle down on you, but *honu* are not likely to bite. Besides, almost any act of resistance will probably drive the *honu* off, which is the last thing you want to see happen.

The turtles don't always try to muscle you away. Sometimes they are content to share their space. The *honu* glides in and plops down right next to you. This is one of the greatest thrills, because it means that the turtle has decided that you aren't a threat. Your obligation is to repay that trust by respecting the turtle. Look, but don't touch. When you move, move away from the turtle. Reinforce the idea that you're not harmful.

Honu, like most wild animals, take their cue from their companions. If you come upon a group of turtles resting together and one is startled into leaving, chances are that the herd instinct will take over and others will do likewise. On the other hand, if one of them is a tolerant turtle and shows no concern about your presence, the others notice and are more likely to stay.

Setting a good example is important when a turtle tries to bully you. The *honu* that is trying to intimidate is obviously unafraid, but what of other turtles that might be watching? If you show that you'll yield to a turtle, they'll gradually lose their fear of you. Eventually, you'll be able to get a much closer look.

One last point to keep in mind: Real etiquette means that you pay attention, learn which spots *honu* prefer, and don't squat in them. After all, it's their reef.

Disclaimer

Over the years, we've been privileged to witness countless turtle interactions. The problem is that we really don't know *why honu* act the way they do. We can tell you *what* they do, but we can only guess at the why. Sometimes their motivation seems obvious, such as the classic turtle turf

Makana, Hawaiian for "gift," usually occupied the same spot underneath a coral ledge on a reef finger that we soon began calling Makana's Ridge. On this occasion, the little juvenile had sat patiently for several pictures but eventually decided that enough was enough. This flipper swipe told us that the shoot was over.

tussle, but other actions look almost random. Perhaps they are — or perhaps *honu* have motives we just can't comprehend or weren't around to witness.

At any rate, the descriptions that follow are faithful reports of what we've seen *honu* doing, but we can only speculate why they are doing it. We thought that it is important to point this out.

Honu body language

We can't be sure that *honu* communicate to each other by the way they position themselves and the motions they make, but after seeing various postures and actions repeated so often, we find it hard to conclude otherwise. What follows are examples of "Turtle Talk" — a kind of *honu* body language — and what we think they mean.

Note that while these are *honu*-to-*honu* signals, we've seen them used on humans. Remember, once the turtles accept you as nonthreatening, they'll treat you much the same as they treat each other.

THE FLIPPER SWIPE. If *honu* signals were mountains, this one would be Mauna Kea, the Big Island volcano that's taller than Mount Everest

Wana, having just returned from taking a breath at the surface, asserts a claim to her spot by summoning up a yawn. In particular, the serrated lower jaw makes it an intimidating display. The *honu* are not likely to bite you, however.

from base to summit. It's the most important signal to recognize. The *honu* extends a flipper and brushes it briskly down from the eyebrow and past the cheek. It can be a casual swipe, done just once, or it can be a vigorous movement that is often accompanied by exaggerated neck and head motions.

The flipper swipe is a definite sign of irritation and serves as a rebuke: "Go away, you bother me." We've seen *honu* make this gesture toward other turtles, divers, and even fish. It's a clear and strong proclamation of annoyance, the equivalent of a human shaking a fist. There is no ambiguity in meaning here — if you see it, you should be immediately on your way.

THE YAWN. Turtles can't really yawn underwater, but they seem to do it anyway. They extend the neck, gradually opening the mouth to a gape and then slowly closing it. This is exactly the same as for a yawn, but of

If a *honu* has found a place to rest peacefully, there's a good chance another turtle will try to take the spot away. Although no other *honu* is in sight, an arriving turtle often decides that the only place to rest is the one that's already occupied. These pictures illustrate the typical spot-usurping strategy as exercised against both *honu* and human: Approach from behind and nip the rear flipper.

course no air is involved. Sometimes the *honu* holds the gape while gazing purposefully around.

The yawn seems to be a territorial signal. You'll see turtles do it most often immediately after they land. We think that in those circumstances the turtle is announcing, "I'm here now and I plan to stay."

The other common use of the yawn occurs when another turtle (or diver, for that matter) approaches. In this case, it seems to mean, "I feel crowded. Back off!" Naturally, you will.

Between turtles, an unheeded yawn can turn into a squabble. One turtle turns the yawn into a snap at the other. Usually the *honu* with the bigger gape wins. This is because a big yawn usually has a *big* turtle behind it.

THE NIP. On occasion, one turtle will nip lightly at another, usually on the trailing edge of a hind flipper. It seems to be more of a way to make a point rather than an attempt to injure. We've noticed that a nipping *honu* generally wants another turtle to move—a claim-jumping tactic, in other words.

We've had over 2,500 scuba dives with the *honu,* and perhaps half a dozen times a *honu* has nibbled at one of our flippers. We see those incidents as examples of the way tolerant *honu* treat divers as other *honu.* It's not a frightening experience at all; in fact, it's rather cute.

THE HEADBUTT. The headbutt — or if it's done from behind, the "goose" — is clearly another way to grab territory. The head goes down, the neck is drawn into a tuck, and the *honu* runs into the other turtle and shoves. This usually results in the victim moving, either voluntarily or not, and a new land claim for the perpetrator.

Occasionally the original occupier refuses to move or even contests the attempt. More often, the *honu* simply shifts over or leaves with a few flipper swipes toward the invader.

If you've earned the *honu*'s trust, and if you lie quietly watching them near prime *honu* resting terrain, you might well be subject to a headbutt — or maybe you'll be startled by a goose. In either case, there's nothing to fear. There's no sudden impact. You'll abruptly become aware that something big is trying to move you — but surprisingly gently. Of course you'll give the *honu* the space. After all, you are loitering near their bedroom.

THE TUCK. A turtle feeling crowded or imposed upon might show it with this posture: a subtle drawing in of the head, often with front flippers tucked under the plastron. The turtle seems to be trying hard to look grumpy. The result is a genuine turtleneck, just like the sweater.

This posture indicates that you've unwittingly gotten too close. Usually, backing away and allowing just a bit more space is enough to reassure the turtle and put it in a more social mood.

THE SPRAWL. This posture really isn't a signal, it's more of a statement. The meaning should be obvious to any couch potato. The turtle is completely comfortable and laid back. The translation is, "My universe

Another classic claim-jumping tactic is in progress. The *honu* in the foreground is about to be rudely butted from behind. On this occasion, the victim meekly surrendered the spot without a fuss or any sign of displeasure. *Honu* are typically mild mannered and easygoing.

is unfolding as it should." Like content humans, this is the mood a turtle seems to prefer. Remember that your approach, especially when made carelessly, can change that mood.

A sprawling *honu* is at rest. A turtle at rest prefers to stay at rest until the time comes to breathe. Few creatures like to remain at rest as much as Hawaiian green turtles do. They're great at doing nothing all day. They don't hang out "Do Not Disturb" signs, but they shouldn't need to. Re-

It would be hard to find a turtle more laid back than Amuala. That's one of the reasons we chose him as a subject for George Balazs' studies. He's also provided us with numerous photo opportunities. On this particular shoot, he'd allowed several photographs but finally had enough. He showed us the tuck. He retracted his neck, hunched his head down, and just generally tightened his body to signal that the shoot was over. This is a *honu* "grump," and a respectful diver will leave quietly.

member, making them move when they don't want to is a prime violation of *honu* etiquette and the Spirit of Aloha — not to mention the law.

The *honu's* castle

Your home might be your castle, but if you're a *honu*, your castle is where you like to rest. The nature of the *honu* is such that where they rest is the same place that they rested yesterday — and the day before that, the month before that, and probably even the year before that. In other words, they have what scientists call "site fidelity to underwater refugia,"

One of the reasons *honu* are so amusing to watch is their obvious delight in just lying around, even in awkward positions you'd think were uncomfortable. Here two *honu* illustrate the posture we call the sprawl, conveying a tranquil mood that we find equally relaxing.

which is a fancy way of saying that they are extremely faithful to specific spots on the reef.

It has long been accepted that *honu* — and sea turtles in general — prefer to remain in the same small area. Even before scientific studies provided confirmation, there were numerous anecdotes about turtles with strong site fidelity. For example, fishermen reported accidentally catching the same turtle repeatedly at the same spot. Even when the "nuisance" *honu* was transported and released many miles away, it would not be long before the fishermen would pull up their nets only to find that their nemesis had returned.

After we had been diving with the *honu* for a while, we realized that this site preference was a lot more specific than anything we'd found reported. We noticed that the same turtles rested in *exactly the same place on the reef* — year after year after year! In fact, the location where we see a *honu* turns out to be a reasonably reliable way to know who that *honu* is. It's not perfect, of course, so we don't depend on this technique for anything but a sense of who's around, but it is still a useful thing to know.

Perhaps the most impressive turtles in this respect are the ones that

Another joy of long-term observation: In the late 1980s there weren't many turtles at Hono-kōwai. We'd seen fewer than a dozen *honu* before we met this turtle in 1990. She carried tags and was the largest *honu* we'd ever seen. She felt so old to us that we called her Tutu, Hawaiian for grandmother. How wrong we were. Later we discovered that Tutu was a young adult female back then. Her first recorded nesting season was just two summers before. Every summer for the next fifteen years, either we saw Tutu or the monitors at French Frigate Shoals did. This photo is one of a series, taken as she took advantage of a pointed coral mound for a plastron scratching session. The worn coral shows how much Tutu and other *honu* use this location.

migrate to the nesting grounds and then come back to settle into precisely the spot on the reef that they'd left months earlier.

Some Honokōwai females have external tags. In other cases, we've recorded the external temporary marking made at the French Frigate Shoals and reported it to George Balazs. Every summer, we compile a list of the females that we've identified and that we expect to migrate. The fine folks monitoring the nesting beach keep watch for these *honu*, verifying that they've nested. We then wait patiently for their return.

Tutu (U521), Shredder (A240), Mendelbrot (U359), Tiamat (122C), Puʻi-

puʻi (U249), Lomi (U164), and Raphael (PIT tag) have all made the risky journey, most of them several times. They all have reliably returned to the reefs of Honokōwai after each trip. We can count on Tutu and Tiamat in particular to snuggle down into precisely the places that they left several months before.

The faithfulness of sea turtles to their nesting beaches is legendary. Based on our observations at Honokōwai, the faithfulness of the *honu* to their special spots on the reef is equally remarkable.

7 Those Other Guys

. . . .

Out of discord comes the fairest harmony.
— Heraclitus

ʻEa

While this is a book about *honu*, we would be remiss if we didn't include an introduction to the *honu*'s cousin Hawaiian sea turtle: the hawksbill, *Eretmochelys imbricate.* Some Hawaiians call them *honuʻea*, others use the name *ʻea*. The definitive Hawaiian dictionary by Pukui and Elbert (University of Hawaiʻi Press) lists both names without stating a preference.

Although *ʻea* has other meanings, every authority we consulted agreed that it is accepted as a name for the hawksbill. On the other hand, some rejected *honuʻea*, saying that *honu* and *ʻea* are two different species and their names should not be combined. We've chosen to use the shorter name, but there is an interesting exception that we'll get to later in this chapter.

ʻEa are much less common than *honu* and are classified as an endangered species under the U.S. Endangered Species Act. The chances that you will see an *ʻea* are low — but not zero by any means.

Hawksbills in Hawaiʻi don't migrate long distances to reproduce. They nest in the main islands, primarily on the Big Island, with some nests reported on Maui, Molokaʻi, and Oʻahu. Monitoring teams protect the known nests on the Big Island and Maui. They do everything possible to ensure that more hatchlings get into the water so that the population of *ʻea* will grow.

For example, they've posted "Turtle Crossing" signs on the highway into Kīhei, Maui, and erected fencing between the road and the shoreline. That's because *ʻea* nest along that coast and have been known to crawl beyond the beach and onto the road that runs close by — too close for the safety of hawksbill mothers.

The conservation efforts seem to be working, because we've documented eight *'ea* at our dive sites. Like *honu,* you can identify hawksbills by their faces, so we keep track of *'ea* in the same way we do *honu.*

The face and head are actually the most obvious means to tell *'ea* from *honu.* As the name "hawksbill" suggests, they have a distinctive pointed beak. Their serrated shells, so beautiful when polished, are actually duller and less spectacular than *honu* shells underwater. The scutes of the *'ea* carapace are imbricated, meaning that they are overlapping. This is often obvious at the hind end, so if you see what look like prominent "cracks" in the rear of the shell, right above the tail, you are probably looking at a hawksbill.

The flippers of the hawksbill provide another way to identify them. *'Ea* flippers have two claws on each compared to the *honu'*s single claw. Their flipper scales are outlined in white more clearly than those of *honu.* The flippers are tiny compared to a *honu'*s, especially in relation to their body size. In general, *honu* are larger in all respects.

Then there are those eyes. *Honu* eyes seem docile, languid, even bovine. They're herbivore eyes. A hawksbill's eyes, even during resting, remain alert — an underwater radar system. *'Ea* eyes seem more like carnivore eyes: I-must-hunt-for-my-food eyes.

Despite all these differences, from a distance even an experienced turtle watcher can mistake a really young *honu* for an *'ea.* A few moments of observation will usually settle the issue, however, especially if you focus your attention on the head.

'Ea differ from *honu* in more ways than just appearance. *Honu* mostly like to spend their days lying around on the reef, usually in the company of other *honu. 'Ea* feed during the day and are more likely to be seen alone and on the move. The normal foraging habitat for *'ea* is the shallow fringing reef in the nearshore waters of all the islands — the resting part of the *honu'*s *kuleana.*

'Ea prefer to eat sponges that grow down between and under the corals. They therefore have to work hard for their food. If you are lucky enough to spot an *'ea,* there's a good chance that it will be tearing up the reef to get at some sponge. Holes with newly broken and exposed coral are a sure sign that a hawksbill is somewhere about.

Like their *honu* cousins, *'ea* are tough on the corals — just in a different way. They also seem to be tough on *honu.* When we've seen them resting together — it's rare, but it happens — the hawksbill and *honu* tolerate each other without fuss. Once the *'ea* is moving, however, it's clear who has the right-of-way. Even the largest *honu* scoots out of an approaching hawksbill's path.

While we're resting on the bottom watching a dozen green turtles, there's an air of peace and tranquility. *Honu* trips to the surface and back are lessons in graceful slow motion. Your blood pressure can dive twenty points in the company of resting *honu*.

When a hawksbill approaches one or more contented resting *honu*, it's much like dropping a squirrel into a tea party. Something will happen, and you can bet that it will involve intimidated *honu* and a hawksbill perpetrator. *'Ea* have sharp, pointed beaks, and they use them without hesitation.

Since hawksbills are thought to be so scarce, we aren't sure why we've been blessed with so many. Possibly *honu* attract *'ea* in a roundabout way. We have a roundabout theory.

Extracting sponges from corals is difficult work. The hawksbill must first find a sponge and then dig down, often plucking out coral piece by piece and flinging it aside.

Imagine tiny underwater backhoes at work. That's the foraging *'ea*. The strength in their necks is impressive as they strain to pull up a coral chunk, working ever deeper, finally gaining access to the prize they know is there.

Now contrast this to the much easier foraging of *honu*, whose food grows on the rocks out in the open. It's much like the difference between the pecking life of a barnyard chicken and the dig-dig-dig of a badger.

That's the point. When you dig for your daily living, it's to your advantage to forage in areas already "excavated." A *honu kuleana*, with its Turtle Tramples and cleaved corals, provides great places to prospect for sponges. The *honu* arrivals and departures of each new day expose more coral, bringing a hawksbill that much closer to its food.

We therefore speculate that *'ea* benefit from the presence of large numbers of their bigger and heavier *honu* cousins. So it's possible that if you come upon a well-worn *kuleana*, a hawksbill will show up sooner or later.

We used to think that *'ea* are just bad tempered. That's because when we've seen a hawksbill swimming past a *honu*, the *'ea* has usually veered and rousted the *honu*. If our theory is correct, however, there's another explanation: The *'ea* simply wants the *honu* to move. There could be an easy meal underneath.

Ake (Hawaiian for "Archie") was only the second Hawaiian hawksbill we'd ever seen. This photo is from our first encounter in 1999. Ake was busy breaking and ripping away at corals, trying to get at the black sponges that she knew grew there. We were a bit surprised at how blasé she was around divers, until we learned that Ake makes her home at a popular site known as Kahekili (Airport Beach) just to the south of Honokōwai. This means that she sees snorkelers and divers regularly. In December 2005 we received a report that a diver had spotted Ake and read her brand new tags. It turned out that she had been tagged on July 16, 2005, at Pōhue on the island of Hawai'i, where she nested three times during her first nesting season.

Other strangers

Over the years, we've met two turtles that appeared odd. There was something not quite *"honu"* about them. Both demanded closer attention.

Ho'omalu

The first oddity was a turtle whose carapace and face looked subtly different. Her carapace was a bluish gray, with a speckled pattern we don't see on *honu*. The beak was slightly rounder, and her eyes were somehow not quite *honu* eyes.

Then there was the plastron. This was what really made her stand out from *honu*. Unlike a *honu*'s yellowish-orange hue, this turtle's plastron was a definite gray.

We concluded from this detail that Hoʻomalu must be either partly or completely a Pacific black turtle, the name often given to the green turtle population of the eastern Pacific. (Some experts argue that black turtles are a separate species.)

We named the turtle Hoʻomalu, a Hawaiian word with several meanings, but the sense we intended was "to bring under the care and protection of." We were certain that her presence at Honokōwai was just temporary, so we hoped that the name would keep her safe throughout her transpacific journeys. We reported the turtle to George Balazs, who took an immediate interest.

George eventually met Hoʻomalu while diving with us. Because she was calm and tolerant, he was able to approach her and achieve a unique method of DNA sampling: Without any apparent discomfort to Hoʻomalu, he gently plucked a few skin barnacles.

Enough shedding skin clung to the barnacles to allow DNA testing, and when the results came back our suspicions were confirmed. One of Hoʻomalu's parents was a *honu*, but the other was a Pacific black.

There was another surprise in store: We continued to sight Hoʻomalu in subsequent summers. It turns out that this turtle calls Honokōwai home.

In an unusual turn of events, a *honu* picks a fight with a hawksbill. Usually *honu* "respect" *ʻea*. Not this time. The passing *honu* noticed the peacefully foraging *ʻea*, then deliberately turned, approached, and snapped. This incident is unique in our experience.

Wai? (top, swimming) and Hoʻomalu (bottom, resting) are both examples of extremely rare cross-breeding: Wai? across species (*honu/ʻea*) and Hoʻomalu across populations (Mexican/Hawaiian). To encounter an example of either is highly unlikely, but to find both of them in the same small patch of Hawaiian ocean tests believability—yet there they are.

Wai?

Another turtle with a different look caught our attention in 2004. From a distance, we first thought we were seeing an *ʻea*. The face had a hawk's bill. Seen in silhouette, the turtle looked and moved like a hawksbill. The front flippers featured boldly outlined patterns like those of an *ʻea*. The rear of the shell was serrated and some scutes were slightly separated, which are *ʻea* attributes.

When we got closer, however, we noticed many *honu* characteristics. Apart from the rear of the carapace, the shell was definitely that of a green turtle. There were only two prefrontal scales (the ones between the eyes), just like a *honu*. A hawksbill has four.

At first, we could see only one claw per flipper, again like a *honu*, but a close examination of our photographs revealed that this odd turtle has vestigial claws on the front flippers just where you'd find a hawksbill's second claw. Here we had a creature that didn't seem able to decide which species it wanted to be!

So we named this turtle "Wai?" — complete with question mark — which is the Hawaiian expression for "Who?" Just who was this?

We sent over a dozen pictures of Wai? to George Balazs, who agreed that this odd fellow could be a cross between a *honu* and a hawksbill — a "honubill," he dubbed it. George eventually got to see Wai? for himself.

Here you can see Wai?'s fully grown claw at the position where a *honu*'s single claw is found and a partially emerged claw in the foreground exactly where an 'ea would have a second claw. This vestigial claw is strong supporting evidence that Wai? is a crossbreed.

He attempted the same DNA sampling method that he pioneered with Hoʻomalu. Unfortunately, this time the turtle's barnacle samples didn't have enough skin adhering to them for the lab to conduct tests.

Waiʔ still calls the reefs of West Maui home. George might yet get enough DNA to confirm that we really have discovered a honubill — a turtle with a legitimate claim to the Hawaiian name *honuʻea,* but like George, we're already convinced that Waiʔ is a true hybrid.

Hoʻomalu and Waiʔ represent yet another facet of Hawaiian turtles. Meeting them reminded us why the *honu* are so fascinating. Each dive offers another opportunity to make new discoveries, and the Hawaiian ocean serves up plenty of surprises.

8 *Honu* Relationships

. . . .

"You're very clever, young man,
very clever," said the old lady.
"But it's turtles all the way down!"
— Stephen Hawking,
    A Brief History of Time

Honu and ancient Hawaiians

No one knows when *honu* came to Hawai'i, but they probably began to show up soon after the oldest islands were formed. Today, more than five million years later, most turtles have settled into foraging grounds around the eight main islands.

For almost all of their history, *honu* lived without human contact at all. It's been just a thousand years or less since the arrival of the Polynesians who would become the first Hawaiians.

Turtles were common throughout Polynesia, so these early settlers would not have been surprised that *honu* were already there. This almost certainly pleased the new Hawaiians, who knew how to make use of the turtles.

Kalei Tsuha, culture and education coordinator for the Kaho'olawe Reserve Commission, gave us interesting insight into the way the early Hawaiians might have thought about *honu*. She says that if you could ask these ancients to describe their relationship with the turtles, the answer would probably be a single word: "*Ono!*" That's Hawaiian for "delicious." To them, *honu* were primarily a prized food source.

Although *honu* meat is nutritious and is said to be delicious, it was not a staple of the Hawaiian diet. They hunted turtles, but they didn't take large numbers. In fact, for some time Hawaiians observed a tradition that only the men of the *ali'i* class, the Hawaiian rulers, could eat *honu*. It was *kapu* (forbidden) to everyone else. Eventually, and particularly after the arrival of the Europeans, this prohibition died out. Anyone could partake.

No matter who ate the meat, the Hawaiians put other parts of the turtle to good use as raw materials. According to Kalei, they made decorative as well as functional items from *honu*. They fashioned the bones from the plastron into sewing needles, and they turned the scutes into fishing lures and ornamental items. They used the rib bones of the carapace to scrape *wauke* and *olonā* (Hawaiian shrubs) in order to make *kapa* (cloth) and rope. Turtle oil served as a sunblock. They wasted nothing.

The relationship with the turtles wasn't entirely material. Many Hawaiians also felt a spiritual connection with *honu*. They honored certain *honu* as their *'aumakua*. These are personal or family gods — venerated ancestors who could take the shape of a *honu*. Some Hawaiians still respect this bond, but of course this relationship is personal. That means it's not generally discussed or written about.

When we began this book, we thought we would be able to find a rich record of the role that turtles played in Hawaiian culture. We did find many references confirming that *honu* were indeed important — but that was all. None of them provided details. This struck us as strange, but gradually we came to realize certain things.

First, the religious aspect is personal or family related. In other words, it is private. We can understand and respect that.

Second, for most of their history the Hawaiians had an oral tradition. They passed their history, legends, and customs from generation to generation through chants and the spoken word. Although the missionaries introduced writing in the nineteenth century, the oral tradition remains strong even today. Many things have never been written down, and much of the written material has never been translated from Hawaiian. This makes it difficult to do research into something like the *honu* in Hawaiian culture.

Finally, we discovered that it is a mistake to assume that Hawaiian culture is homogenous. Each island has its own traditions and legends. Some are shared, but it's not uncommon to find variations, even from family to family.

Still, there's ample evidence that *honu* have always played a role in Hawaiian culture. Blaine Kamalani Kia, cultural adviser to the annual Aloha Festivals, says that the *honu* is the Hawaiian symbol of longevity, peace, humility, and the spirit within.

Honu inspired a turtle chant and a hula in which the dancers imitate the motions of a turtle. Turtles also turn up in other chants and hulas.

Honu also appear in Hawaiian art, including the oldest form: petroglyphs. Not many of these remain, but the few that do include representations of the turtle.

All of these things indicate that the relationship of the Hawaiians to *honu* was one of respect, appreciation, and in some cases even veneration. Then, in the nineteenth century, whalers began coming to Hawai'i. Their relationship with the turtles was drastically different.

Honu in the era of exploitation

The Hawaiians were no threat to the *honu* population. The *honu kapu* resulted in a limited harvest and careful use. The *kapu* meant little to the new arrivals, however. They had no reservations about hunting the turtles, especially in the remote Mokupāpapa, where basking turtles were easily captured. Their relationship with *honu* was the same as it was with the whales: They were prey.

In those days, anyone who spent months at sea knew that green turtles were a useful resource. Because turtles could stay alive a long time in the hold of a ship, they were a prized source of fresh meat. Sailors took green turtles whenever they could, and Hawai'i was no exception.

For a long time, this harvest was not a danger to the population, but the count of *honu* taken kept increasing. When air travel eventually turned Hawai'i into a popular tourist destination, no one anticipated the unintended but nonetheless disastrous effect on *honu*.

Quite simply, the visitors found turtle steaks to be just as tasty as the Hawaiians did. Restaurants bought more and more *honu* at a handsome price. Modern advances such as motorboats, synthetic nets, scuba, and spearguns made it possible to catch turtles in large numbers. In the foraging grounds, hunting gradually increased until there was a substantial commercial harvest in the 1960s and early '70s.

Meanwhile, the turtle population suffered double damage at the nesting grounds in the French Frigate Shoals. Despite the isolated location, turtle hunters knew where to find the nesting *honu*. Since turtle eggs were considered just as much a delicacy as turtle steak, it was inevitable that nests would be dug up and emptied.

Plundered nests weren't the worst of it, though. Since basking and nest-

These *honu* are lucky. Before 1974, a pile of turtles in a fishing boat would have meant they were on their way to slaughter. This picture is from 2005, however, and these turtles had been captured for research purposes. A short time later, they'd all been tagged, measured, assessed for health, and placed gently back in the water.

ing *honu* were helpless once out of the water, they made easy prey. Hunters could simply walk up to a turtle and carry her off, thereby removing a breeding mother permanently from the population. We cannot begin to calculate the number of future generations lost in this way.

At the time no one knew how many *honu* there were, but clearly this sort of harvest could not be sustained for long. Fortunately for the turtles, their relationship with humans was about to change again.

Honu in the era of protection

In 1969, George Balazs was a young biologist who had yet to choose his field of study. While he and his wife Linda were watching a Maui fishing boat unload a catch of *honu* for delivery to local restaurants, he began to wonder how many turtles were left out there. This curiosity led to a notable career in which George and his colleagues have made huge contributions to what is known about *honu*.

One major early discovery confirmed that the number of *honu* had dwindled dramatically: In 1973, George counted just sixty-seven nesting females at East Island in the French Frigate Shoals. Their population had gotten so small that the continued survival of *honu* was in doubt. He began to fear that there was a genuine danger that the last turtles could wind up as steaks on the plates of tourists.

George began working to protect and study *honu*. In due course, his efforts and those of others began to turn things around for the turtles. The passage of a 1974 Hawai'i state law protecting the turtles, coupled with the listing of *honu* as "threatened" under the U.S. Endangered Species Act in 1978, fostered the recovery of the population. The story is told in detail in Osha Gray Davidson's powerful book, *Fire in the Turtle House* (PublicAffairs, 2001).

In fact, without the work of George Balazs, we wouldn't have written this book because it's likely there would be far fewer Hawaiian turtles — maybe none — and certainly not enough to make turtle watching practical. Whenever you see one of the beautiful creatures called *honu*, you can thank George and the many people of Hawai'i who have spoken out and acted on behalf of the turtles over the past thirty-five years.

Honu today

Honu's return from that low point in their history is nothing short of remarkable. There are few conservation success stories to match it. The people of Hawai'i have every right to be proud of the recovery of their *honu* population.

Today, the relationship between *honu* and humans is twofold: institutional — involving various government and nongovernment groups that are working to further protect *honu* and educate the public; and personal — the one-on-one relationships that are formed when people and *honu* meet in the water and on the beaches. As you will see, the former stimulates the latter.

Honu Ambassadors

Sea Life Park on O'ahu has played an interesting role in the *honu* recovery. In 1989, Sea Life Park's former curator Steve Kaiser started the "Hawaiian Sea Turtle Ambassador Program." This was a visionary conservation plan in which *honu* actually did service on behalf of all marine turtles.

Steve knew that little *honu* introduced to the public would be their own best representatives. Sea Life Park had obtained several large *honu* in 1968, before laws protecting them had even been conceived. These turtles began nesting on Sea Life Park's artificial beach in 1976, with most of the resulting hatchlings — from a few dozen to a few hundred each year — released into the wild.

Steve's plan called for lending some of these hatchlings to qualified aquaria across the United States. When the loaned turtles grew to about fifteen inches, the hosts returned them to Sea Life Park. There, after a thorough health check, they were ultimately released into Hawaiian waters.

These young *honu*, on public display around North America, became symbols for all marine turtles to hundreds of thousands and perhaps millions of people. Since little *honu* are irresistibly cute, they undoubtedly advanced the cause of sea turtle conservation enormously.

Although the original Ambassador Program ended in 2001, a limited variation of it still exists. Sea Life Park still sends some *honu* to be raised in properly maintained environments where the public can get to see them.

Today in Hawai'i, you can see Sea Life Park Ambassadors at the Maui Ocean Center, where the research projects include studying the captive diet of young *honu*. The Hilton Waikoloa Village on the Big Island features little *honu* Ambassadors in its ocean-fed lagoon.

Honu Ambassadors also charm guests at the Mauna Lani Bay Hotel on the Big Island. There, a series of beautiful interconnected pools hold several tiny *honu* that are getting ready to participate in the annual Turtle Independence Day, celebrated each Fourth of July. When the big day arrives, crowds as large as a thousand people gather to see the elaborate and moving ceremony that accompanies the release of each *honu* youngster into the bay.

Showing turtles aloha

Honu prompted the creation of another public education program by simply crawling out of the ocean to bask. This might not have required action had the turtles chosen to do this somewhere other than the beach at Laniākea on the North Shore of O'ahu, right next to the busy Kamehameha Highway.

Laniākea is world famous for its surf break. Since 1999, it has also become known as a place where you have an excellent chance to see one or more *honu* snoozing on the beach — provided you can find a place to park.

Laniākea had always been busy because it is a popular surfing spot, but the basking turtles began attracting increasingly larger crowds. Most visitors respected the turtles and admired them from a distance. Some, however, couldn't resist getting too close and even touching a *honu*. People even straddled the turtles to pose for pictures.

Something had to be done, so in 2005 George Balazs created the "Show Turtles Aloha" campaign, designed to educate people about the basking *honu* and how to behave around them. The idea was to maximize the turtle watcher's experience while minimizing the effects on the turtles.

George placed informative banners and signs around the beach and enlisted volunteers. These "*Honu* Guardians" handed out literature and answered questions. When a *honu* hauled out to bask, they placed a red rope at a discreet distance around the turtle so that people would know how close they could approach.

Today the campaign carries on under the combined management of NOAA's Pacific Island Regional Office and a volunteer coordinator. *Honu* continue to crawl ashore, in the process engaging the public and winning friends from near and far.

Making nests

There is yet another way that some people are lucky enough to form a relationship with *honu:* on the nesting beach — and not in the remote, protected French Frigate Shoals but in the main islands. This is a recent development.

Within recorded Hawaiian history, there is only one well-documented main-island nesting site for *honu:* Polihua, a beach on the northern shore of the island of Lānaʻi. (The name translates literally to "eggs in bosom.") Hawaiian folklore describes how ʻAiʻai, the fish demigod, marked a stone at nearby Kaʻena, the northwestern point of Lānaʻi. The stone then turned into the first *honu*. This legend is the Hawaiian explanation for why the turtles came to Polihua to nest.

Honu stopped returning to Polihua decades ago, but now they have started to nest in the main islands again. Although it is still a rare event, the *honu* population recovery is slowly increasing the chance of seeing a turtle crawl up on a beach — not to bask, but to lay eggs. As already noted, 5690 has been nesting on a Maui beach in Lahaina every other year since

2000. Reports of other green turtle nests are starting to trickle in. Since sea turtles return to lay their eggs on the beach where they hatched, in a couple of decades their hatchlings will return. Eventually, nesting in the main islands could become commonplace.

We certainly hope so. There is no other experience quite like watching a *honu* mother making her nest. If you are lucky enough to witness such a thing, you probably won't be able to forget it. Quite likely you will bond with the turtles forever, if you haven't already.

Individual contact

Honu have a remarkable charisma that charms almost everyone who encounters them. People simply fall in love with the turtles.

Over the years, we've met several people who once killed *honu* for a living. Like many hunters, they felt a special bond with the animals they pursued. Without exception, all of them continued a relationship with *honu* after protection was introduced: They all became turtle advocates. While the recovery of the *honu* population has triggered a call from some Hawaiians to allow turtle hunting again, none of the former *honu* hunters we know supports that position. They've come to cherish and value the turtles and understand better than most what they mean to Hawai'i and its people.

Our website (Turtle Trax, http://www.turtles.org) has prompted numerous people to write to us describing their experiences with *honu*. The typical reaction is identical to ours when we first saw Clothahump: love at first sight. Often there is a spiritual connection. Somehow the turtles touch the souls of those they encounter. This often is not easily described, but those who have experienced it just *know* it has happened. It's somewhere between falling in love and having a religious revelation, but it's not either of these, really. It's the Spirit of Aloha for *honu* entering your heart, and it's unforgettable.

The most touching, heart-breaking turtle relationships shared with us are tragically brief. They begin with a contact from one of those organizations dedicated to granting the wishes of children suffering from a fatal disease. A child's cherished desire turns out to be swimming with or just seeing sea turtles. Fortunately, *honu* are ready to oblige. They are easily accessible at Laniākea even to those who need assistance with mobility. Nowhere else in the world is there a better opportunity to satisfy this simple yet seemingly difficult request.

Our personal experience

Our own relationship with *honu* began with Clothahump. It was immediately one of fascination, admiration, and infatuation. While these elements have never varied for us, as we came across more *honu* another factor eventually dominated the relationship: tumors. We saw the turtles as individuals, each with a story. We watched helplessly as virtually every *honu* developed tumors, one by one. Many disappeared.

We struggled with this for many years. While the suffering of the turtles ripped at our hearts and was the source of much grief, there were compensations. If not for *honu* and their plight, we would never have met many wonderful people who share our love for the turtles, from scientists to enthusiasts like ourselves.

Along the way, we learned enormous amounts about the turtles, about marine science and the scientific method, and about people. Best of all, we were able to contribute in a small way to the realization that the tumors, while still a potentially fatal torment for individual *honu*, have not placed the entire population in peril.

9 *Honu* in Distress

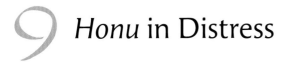

The good news

Honu are the most fortunate sea turtles on the planet. Their nesting grounds are remote and uninhabited. Further, the area has long been a National Wildlife Refuge managed by the U.S. Fish and Wildlife Service, and it has now been declared a national monument. Even if you could get there, which is difficult to do, you are not allowed to enter without a permit.

Consequently, *honu* aren't bothered while nesting. Their eggs are safe from poaching. Only other *honu* or large waves can disturb the nests, and there are no human beach activities that could affect them. While Hawaiian monk seals and nesting turtles sometimes crawl over or rest on nests, they don't pack the beach so hard that hatchlings can't dig out. When hatchlings do emerge, there are no artificial lights to distract them in their scramble down the beach into the water.

The U.S. Endangered Species Act and the wildlife laws of the State of Hawaiʻi have protected the turtles for decades, both at the nesting grounds and in their *kuleana*. Since the United States enforces the laws protecting endangered and threatened species, *honu* are blessed with habitat that is just about as safe as it can be.

Still, threats exist.

Hatchlings as prey

Many natural predators are capable of swallowing a bite-sized *honu* hatchling. Crabs are eager to snatch a meal from the stream of tiny, newly hatched *honu* flip-flopping down the beach to the ocean. After they reach the water, hatchlings can be dinner for almost any large fish. Predation in the first few hours of life is the major reason why biologists estimate a hatchling's chance of surviving to adulthood to be less than one in a thousand — and sometimes less than one in ten thousand.

If they get through that first day, however, their chances of survival increase significantly. Their swimming frenzy takes them away from most of the predators. They start to grow. Little *honu* slowly strike themselves off the menu of one carnivorous fish after another, until just one remains. Unfortunately, a *honu* never gets large enough to be disqualified from a large tiger shark's "Top Ten Tasty Things to Eat."

Swimming with sharks

While we've caught only a couple of brief, sobering glimpses of tiger sharks in nearly four thousand hours in the underwater world of Hawai'i, this top-of-the-food-chain predator has left its calling card on many *honu* in the local *'ohana*.

The lucky ones are left with deep white scratches and gouges in their shells but with the rest of the body intact. Sometimes the shell even gets fractured. It's easy to speculate what happens: The shark catches up to the *honu,* grasps with its jaws, and tries to crush the shell.

The shark shakes the turtle and bites down hard. The turtle, helpless, is locked in the tiger shark's massive grip. A lot depends on how much pressure the *honu*'s shell can endure. Should the plates collapse, there goes the whole turtle.

The "slip factor" also comes into play: The smooth shape deflects the teeth so that they tend to scrape down the shell without penetrating. *Honu* with badly scratched up carapaces prove that the shell — and ultimately, the turtle — can triumph.

Partial victories

There's an even more gruesome side to tiger shark encounters: If it can't get the whole turtle, a large tiger frequently contents itself with a body part — usually a flipper.

Imagine for a moment the enormous disadvantage of a missing flipper.

Front flippers supply power for swimming. Turtles can still swim surprisingly fast when missing one, but not as fast as before. Hind flippers are used to steer, so turning is harder without one.

Missing flippers reduce a *honu*'s chance of breeding successfully. First, there's the difficulty of completing a migration with a flipper gone.

Then, when it comes to the act of mating, a missing front flipper essentially disqualifies a male. How, with just one flipper, can he hope to mount a receptive female and hang on?

Females have it nearly as tough. Missing flippers make it far more difficult to nest. *Honu* mothers need front flippers to crawl and dig the body pit, while hind flippers dig the all-important egg chamber. Surprisingly, females with a missing limb *can* nest successfully—a testimony to their amazing ability to survive.

Consequently, sharks reduce the turtle population in more ways than just eating them. Of course, there is no guarantee that because the shark only took a limb, the *honu* will survive. Then again, the number of *honu* with pieces missing in the *'ohana* at our two dive sites confirms that many of those attacked pull through.

We wondered how a *honu* manages to overcome the trauma of having a limb torn away. We knew that shark bites aren't usually immediately fatal, and the greatest risk is death from blood loss. Humans know this and know how to stop blood flow, so many shark-bite victims survive—but what about *honu?* We see enough turtles missing part or all of a flipper to know that somehow they don't normally bleed to death. Why?

We asked Dr. Jeanette Wyneken, author of *The Anatomy of Sea Turtles* (NOAA Technical Memorandum NMFS-SEFSC-470). She explained that part of the reason is that the turtle is underwater. Water pressure slows the bleeding, and the deeper the turtle dives, the higher the pressure.

Another factor she mentioned is that blood in the flippers isn't under high pressure. Most of the turtle's flipper blood goes from high-pressure arteries into a series of much smaller blood vessels, and some goes into a network of capillaries and small vessels before it is distributed to the flippers. Veins are low pressure already, so the overall effect is that a *honu* simply won't bleed as readily as we would.

Finally, Jeanette pointed out that nature has given the turtles some help: They are really good at clotting.

The most dramatic and grievous shark injury we've seen happened to

Our first encounter with George in 2002 resulted in this photo. The shark-bite damage to his right front flipper is disturbingly obvious. The greenish coating is algae growth, indicating that he must have spent a lot of time resting on the bottom recuperating.

a male we've known since 1999. We'd named him "George" in honor of George Balazs.

George had a noble countenance, a larger-than-life presence — and impossibly huge feet. Best of all, he had a splendid shell that shone copper and gold and emerald in his ocean's late afternoon light. Seeing George always gave us great pleasure — until our first sighting in 2002.

When we first saw George that summer, we didn't even recognize him. From a distance we saw only a male turtle in disturbing condition. A coat of algae fouled most of his body. Clearly a large shark had sheared off most of his right front flipper. We could see bone protruding from the wound. Another part of the turtle — we weren't sure exactly what — dangled beneath his chin. Poor, poor *honu!*

When we got to our database and matched the turtle's face, we were shocked to discover that it was George. He was in such horrendous shape that we simply hadn't been able to recognize him as our old friend. We were sure that he was finished. The note in our database for that year simply states, "Expect this year to be his last."

To our surprise and delight, we sighted George again in 2003. He was resting on the reef and looked much better. His wounds had healed and he'd rid himself of the algae that coated his body the year before. He was, understandably, shy of our approach.

We don't know what effect the partial front flipper will have on George. We suspect that he won't have much reproductive success. We wish him the best though — he had one really rough year. He is a superb example of the legendary recuperative powers of sea turtles.

George also illustrates the point that no matter how large *honu* get, there are sharks out there big enough to eat them — or at least parts of them. The good news for *honu* is that once their pelagic stage is over, sharks are their last natural enemies.

Defenses against sharks

As predators, sharks are supreme: splendid streamlining, startling speed, and superb senses combine to create fearsome hunters. Since *honu* are prey to these efficient and ancient foes, they have evolved a number of natural defenses.

The first of these is size. Only big sharks eat big *honu*. To get big, however, *honu* must survive long enough. This is where camouflage helps.

Turtle shells blend extremely well into coral reefs. *Honu* prefer getting right down into any holes around their *kuleana,* making them harder to get at as well as hard to see. *Honu* have truly evolved to blend right into the reef.

Sleeping during the day is another natural defense for *honu*. Sharks like to hunt in the dim light of sunset and dawn, so that's when *honu* have to be awake. The turtles move into shallow water to feed, where they remain alert and hair-trigger wary. Although they tend to stay out to sea, large sharks can and sometimes do come close to shore.

Honu, like all sea turtles, have also evolved to be streamlined for efficient and surprisingly fast swimming. Remember, a determined *honu* can reach speeds upward of twenty miles per hour for short bursts. Although sharks can certainly swim even faster, there is no doubt that the *honu's* ability to accelerate quickly and turn sharply helps them avoid being eaten. A speed spurt might be all the turtle needs to make it to safety under a ledge or into a cave.

If a shark does catch up, the turtle's defensive instinct is to roll and swim on her side, giving the shark the most difficult target possible. She presents the whole carapace to the shark, almost like a shield. Even if

Facing page: Shredder, a female first encountered in 1995, blends right into the coral thanks to her excellent disruptive coloration. Her dappled shell is effective from every angle, a superb example of camouflage.

the shark's mouth can open wide enough, the teeth have less chance of penetrating her shell's slippery curved surface.

So even in the face of such a formidable enemy as the tiger shark, a *honu* is not defenseless. Of course, sharks and turtles have coexisted as predator and prey for millions of years. While tiger sharks can never be discounted as a threat to even the largest of *honu*, they are also a natural predator. The shark has always hunted *honu* and nature has always maintained a balance. Individual turtles are constantly at risk, but the tiger shark is not a threat to survival of the turtle population of Hawai'i.

It's too bad that *honu* haven't had the same opportunity to evolve defenses against us humans.

Humans and *honu*

Until the mid-1970s, the actions of human beings represented the most serious danger to *honu*. We have done — and in some cases continue to do — a number of things that put *honu* in peril, both directly and indirectly.

Hunting

Commercial hunting of *honu* nearly wiped them out by the early 1970s. Fortunately for *honu*, Hawai'i woke up to the turtles' plight and did something about it. Today, killing turtles is illegal and *honu* flourish. Poaching is uncommon, and the number of *honu* has increased to the point where it is reasonable to consider a managed harvest. In the near future, Native Hawaiians could resume the right to catch a limited number of *honu* in accordance with their cultural customs and uses. With proper supervision, such a hunt would not threaten the survival of the *honu* population.

Our affection for the *honu* means that of course we don't want to see them hunted. If, however, their numbers have recovered to the point where a regulated harvest would not threaten the overall population, then we find it hard to oppose. Our objections would be strictly emotional, not scientific. We'd insist that such a hunt be strictly controlled and that safeguards be put in place to prevent hunting in areas where human contact has acclimated the turtles. If these conditions were met, we wouldn't like the idea, but we'd be forced to accept it.

Fishing

An innocent activity that isn't intended to hurt the turtles at all turns out to be the way humans presently affect *honu* the most: fishing.

Scientists have reported that longline fishing occurring far out to sea affects other species of sea turtles — notably the leatherback — but it is not a problem for *honu*. The two kinds of fishing that do have an effect on *honu* are lay gill net fishing in nearshore waters and pole and line fishing from the beach.

There is plenty of lay net fishing in Hawaiian nearshore waters. State laws stipulate that if you set a net, you must check it at least every two hours and release entangled turtles and other nontargeted marine life. While there are occasional *honu* deaths from this fishery, they aren't common and do not pose a significant threat to the population.

Surfcasting and pole fishing from shore are also popular fishing methods in Hawai'i. The danger to *honu* here is not usually death but hooking and entanglement that eventually result in severe injury and, often, the painful loss of a limb.

Fishing line gets wrapped around a flipper and the *honu* struggles with it, pulling it ever tighter. The nonbiodegradable line cuts into the skin until the flipper becomes strangulated, then months later is literally self-amputated. Some entangled *honu* are lucky enough to be reported to NOAA, who then try to find, rescue, and treat the turtle, but many victims simply are forced to endure their nasty fate. A turtle in this predicament not only suffers for a long time but also is seriously handicapped in its chances for long-term survival.

If you fish, always keep in mind that the turtles come right up to shore to feed, usually at dawn and dusk. Please don't cast your fishing line anywhere that you see *honu* surfacing to breathe — and do tell others.

If you do hook or entangle a turtle when fishing, use your dip net or firmly hold the front flippers and shell to lift the turtle out of the water safely. Do *not* pull the line — this causes more injury. If the distance is too great or the turtle is too large, cut the line as short as possible to release the turtle. If you bring the *honu* ashore, cut the line close to the hook. Remove any entangled line. Be wary of the *honu*'s mouth and flippers.

Sometimes Hawaiians use a kite to get a hook far enough from shore to catch big pelagic fish. A large plastic jug serves as a bobber. Unfortunately, this turtle somehow snagged a kite-fishing hook in her flipper. The buoyancy of the jug prevented her from resting comfortably on the bottom. Worse, it impaired her swimming so much that she would have been easy prey for a tiger shark. We know this because it also slowed her enough to allow us to catch up to her and cut the line, resulting in one of our most satisfying *honu* experiences ever.

Don't remove the hook unless you can take it out without further injury. If you're uncertain, leave the hook in place.

Turtles brought ashore with serious cuts or ingested or deeply embedded hooks need veterinary care. Keep the turtle in the shade. Immediately call DOCARE, the State of Hawai'i agency responsible for such

This is not a mounting and mating attempt, although it might look that way. The young *honu* on top is a female, but we didn't know that when this picture was taken in 2003. Apparently she was just bugging Blue (a male) for reasons unknown. In 2006, we found out that she was a female in a most unfortunate way. She had suffered a prop hit that killed her. Her body was recovered, and the subsequent necropsy revealed her gender.

matters. The number is in the phone book, or you can refer to the Web site, http://www.turtles.org/nmfs.

Boats

Naturally, there is a lot of boating activity in Hawaiian nearshore waters — the same waters preferred by *honu* for their *kuleana*. Boat operators are usually aware of *honu*. They are careful in areas where they know the turtles gather, but inevitably there are occasional boat-*honu* collisions.

Honu don't seem intimidated by a boat bearing down on them. Several times we've witnessed turtles head to the surface for air even though we could hear the engine of an approaching boat growing ever louder. *Honu* seemed completely oblivious, or at least uncaring, that a boat was coming. Fortunately, every time we've been watching, the turtle has dived at the last second.

Although it's uncommon, some turtles do get hit, either by the boat's bow (usually not fatal) or by the much more lethal propeller. We can only guess why *honu* put themselves in harm's way like that. They do have poor hearing, but the engines are always loud and certainly create the low frequencies that *honu* can pick up. It seems unlikely that they don't hear boats coming. We're forced to conclude that they just don't understand the threat.

Onshore activity

While people can't cause problems with the nesting beaches, we still do things on land that affect and often can clearly threaten *honu* in the water.

The worst of these is allowing pollution into the ocean. *Honu* are highly faithful to their *kuleana* — even to specific places on the reef. They don't leave just because the water gets dirty. This means that if we allow sewage or agricultural runoff into the water, *honu* stay there and tolerate it.

Again, in a way, *honu* are lucky. Hawai'i doesn't have the kind of industrial contaminants that many other places have, such as toxic chemical disposal. The industries here are primarily agricultural or tourist oriented, such as sugarcane, pineapples, hotels, and golf courses. Still, both kinds have the potential to pollute the ocean.

Agricultural runoff is typically high in phosphates and nitrates from fertilizers. Sometimes it contains pesticides. The runoff from golf courses is often a source of fertilizers too.

Hotels and condominiums sometimes drain their swimming pools into the ocean, but mostly they add to the load on the sewage system. While Hawai'i generally doesn't let raw sewage get into the water, treated effluent almost certainly does. This is sewage that has had harmful bacteria such as *E. coli* removed but has not been cleaned of its phosphates and nitrates. In one sense, it's similar to agricultural runoff with the silt removed. Both are full of things that fertilize.

All this nutrient-high pollution can cause something called *eutrophication* — an overgrowth of algae. While *honu* feed on algae, this abundance is not really a good thing for the ecosystem as a whole. The algae smother the corals and ultimately degrade the reef, eliminating valuable habitat.

Worse, the algae can host potentially harmful microorganisms. One

Runoff from heavy winter rains in 2005–2006 carried red silt right out to the reefs where *honu* spend the day. The red mud collected as deep as six inches in some places. In our experience, it will take years to dissipate since the slightest wave action stirs everything up and makes the water murky again. This soil has been washed from the pineapple fields directly *mauka* (mountainside) of Honokōwai.

type, called *dinoflagellates,* has been implicated in such blights as red tides, toxic shellfish, and ciguatera fish poisoning. For *honu,* there is an even more sinister aspect: Dinoflagellates might spur tumor growth.

FP and *honu*

"FP" is short for *fibropapillomatosis* or fibropapilloma tumors. In the mid-1980s, these tumors began spreading rapidly among green turtles around the world. Inexplicably, they broke out simultaneously in populations in the Atlantic (Florida, the Bahamas, and other places) and the Pacific (in particular, Australia and Hawai'i).

In some years, eutrophication leads to the formation of huge collections of *limu* (in this case, the seaweed is *hypnea,* a nonnative exotic species). We have often found tumored turtles resting in these mats but rarely have seen healthy *honu* doing this. We were never able to determine why this should be. One possibility is that they are hiding—taking advantage of the camouflage the seaweed offers. After all, sick *honu* are more vulnerable to shark attacks.

The sad fact is that in many places around the Islands, when you see *honu* there is a chance that the turtle will have ugly growths that range from barely noticeable to large and grotesque.

Seeing an afflicted turtle will probably disturb you. It certainly upsets us. We were so distressed that we began recording the effects of FP on *honu.* Since 1988, we've identified over 750 turtles, and over 50 percent of them have been affected to some degree.

Over the years, this has given us insight into the disease. We are happy to report that the news is not all bad. Not all *honu* get the disease, and in many cases the tumors regress: They just shrink spontaneously and often disappear over time.

People who see *honu* with tumors are usually saddened and curious. They're troubled to see such a beautiful animal disfigured by horrible lumps. They want to know why it happens and what can be done about it.

Research has shown that the cause is almost certainly linked to a herpesvirus, but it isn't clear how it spreads. Unfortunately, it is all too easy to find well-meaning but simply erroneous information on this point. Web sites, pamphlets, even misinformed divemasters and boat crews will sometimes caution against touching the turtles because this somehow leads to tumors. We've heard everything from suntan lotion to human bacteria cited as the cause. There are excellent reasons not to touch turtles, but FP isn't one of them. Nor is the myth true that somehow humans can catch the disease from the turtles. The people spreading this misinformation are invariably trying to help, but using falsehoods to do it, even unintentionally, is just plain wrong.

Most people ask if pollution is involved. There is a general association between high FP incidence and water problems — usually high levels of phosphates and nitrates from agricultural runoff or treated sewage effluent. There is no direct, provable link, however. Contaminated water doesn't always produce FP, and although it's rare, tumors sometimes break out in water that is otherwise trouble free.

You probably won't see a severe case of FP, but if you do, be prepared for a shock. FP is a hideous disease. Tumors grow primarily on the skin but can also appear between scales and scutes, in the mouth, on the eyes, and on internal organs (which, of course, you can't see). Small at first, they sometimes grow to four inches or more in diameter — the size of a grapefruit.

In severe cases, tumors can obstruct both eyes and even destroy the eyeball, essentially blinding the turtle. Tumors growing in the corners of the mouth make breathing and eating difficult. Tumors *inside* the mouth are tragically unique to *honu*. Frequently, there are growths around the glottis that cause serious breathing problems. Turtles afflicted this badly are typically emaciated and doomed to suffer a slow death. There is no known cure.

In the early 1990s, we thought we were looking at the brink of extinction for *honu*. There were few turtles at Honokōwai, and those we sighted were showing signs of FP.

The beauty of long-term observation is that you get to see how some things turn out. Because we kept records, we now know that an encouraging number of turtles become only slightly afflicted. For some fortunate *honu,* tumors stay limited to the eyes only.

In almost every turtle we've documented with FP, the disease began in the eyes. In those that regressed, the eyes improved first. Because of this, we've concluded that you can get an excellent indication of a *honu*'s FP status by looking at the turtle's eyes.

If there are no eye tumors, this is a strong signal that the turtle is improving even if the *honu* has fairly large body tumors, so take heart. This is only a general guideline, but it does give you a better chance to understand the course of FP in a *honu.*

We learned this by observing and documenting the *'ohana* at our dive site since 1988. Happily, we've seen that most turtles have a good chance to recover — provided that they survive to grow large enough.

In our experience, the smaller and younger turtles are much more vulnerable. If an older, larger turtle gets FP, there is a good chance that the disease will regress within three or four years.

It's a good thing that our dive site isn't typical of the rest of Hawai'i, because this reversal rarely happens to the smallest turtles of the Honokōwai *'ohana.* Instead, the disease progresses more rapidly in them. Fortunately, FP isn't everywhere in the Islands.

In fact, all indications are that FP isn't the dire threat to the population that we once thought it was. Although FP is undeniably loathsome, painful, and definitely can kill, we've seen that the disease frequently isn't fatal. Many *honu* do get better.

Beyond a certain size, tumors surely must cause the turtle considerable discomfort. We humans usually find them repulsive and upsetting to see. Remember, however, that they are by no means a death sentence.

Anyone who sees a turtle with severe FP has a natural urge to do something. Sadly, there isn't much most of us can do unless the *honu* strands ashore or is sick enough to warrant capture *and* is close enough to shore *and* is small enough to handle. Most aren't.

In Florida, the Marathon Sea Turtle Hospital has done a lot of work with surgical removal of tumors, thanks to the skills of volunteer veterinarian Dr. Douglas Mader. While this has contributed greatly to knowledge about the disease, it is not a practical treatment for the whole population. In fact, it isn't practical in many individual cases.

For example, it's difficult to remove eye tumors without blinding the turtle. The tumors that grow inside the mouth and throat of the *honu*

Fibropapilloma tumor disease hits the smallest turtles hardest of all. Like Makana the year before, Akebono came down with FP in 2003. What were just small white spots in 2002 had mushroomed into these tumors ten months later. We never saw Akebono after that summer. We didn't expect to. FP is so depressingly predictable that we no longer name juvenile *honu*. Akebono was the last young turtle at Honokōwai to carry a name.

are almost always inoperable. Even at Marathon, if X-rays reveal internal tumors, they don't attempt surgery. In any case, hospitalization and surgery isn't a practical treatment for an entire population or even a single *'ohana*.

The same problem applies to the other experimental treatments that have been devised — two examples are cryogenics and injectable drugs — and none of them is a broad cure. Besides, there are many reasons to believe that environmental cofactors are at work to cause the disease. Treating individuals cannot solve that sort of problem.

No one is happy with this state of affairs. Sea turtle experts all over the world have devoted much time and effort battling this wretched disease.

George Balazs had joined us on a dive at our alternate observation site in Nāpili, Maui, when a large turtle with a massive tumor swam into view. Essentially, the *honu*'s head was all tumor. The sunken plastron suggested that this turtle was emaciated and slowly starving to death. A turtle in this condition is beyond help.

Nowhere has more been done than in Hawai'i, where George Balazs and his colleagues have made the fight a personal as well as a professional priority.

While a cure or containment measure is not yet in sight, there is good news. Our photo and video records, as well as other studies done in Hawaiian waters, show that FP isn't nearly as deadly as was first feared.

Best of all, the latest analysis shows that FP–related deaths are not affecting the *honu*'s chances for survival in terms of the whole population. *Honu* are growing in numbers throughout their range. Certainly FP kills some, but others live and grow to reproduce and replenish the population. There's comfort in knowing that.

10 Of *Honu* and Foxes

You become responsible, forever,
for what you have tamed.
—Antoine de Saint-Exupéry,
The Little Prince

A perfect allegory

The classic story *The Little Prince* contains a beautiful metaphor for the relationship between humans and nature. In the story, the Little Prince encounters a fox. Feeling unhappy, he asks the fox to play with him. The fox replies that he cannot, for he is not tamed. The Little Prince is unfamiliar with the concept of taming, so he asks the fox to explain.

The fox teaches the Little Prince that taming means establishing ties that, once created, make the bond unique. He points out that the Little Prince is just a boy, like a hundred thousand other boys, and that he is a fox, just like a hundred thousand other foxes. Then he explains that taming would mean that they need each other.

So the fox teaches the Little Prince how to tame him. They gradually get to know one another from a distance of safety—their own comfort zones. Over time, the two hearts draw a little closer every day. Ever so patiently they forge a bond of trust. Taming is time intensive and demands patience. There are no shortcuts.

We hope that you understand what this concept of "taming" has to do with *honu*. There could be a hundred thousand sea turtles in the Pacific that are just like a hundred thousand other sea turtles in the world, but the ones we dive with have tamed us, and we them. They have names like Clothahump, Zeus, Tiamat, and Akebono. Some, like Tutu and Nui, we've known since the early 1990s.

To us, the turtles at our dive site are "unique in all the world." This uniqueness is enormously rewarding. These *honu* matter a great deal to us. We have need of them.

Taming indeed has its emotional rewards, at least for us humans. Yet taming is truly an awe-inspiring thing, if for no other reason than taming teaches a wild creature to trust humans. Taming essentially removes some of the "wild" out of wildlife.

Every single day in Hawai'i, humans are taming *honu*. By not harassing, hurting, or hunting them, residents and visitors alike teach the *honu* that they have nothing to fear from close human contact. As we said before, *honu* learn from experience. Tour boat operators, divemasters, kayakers, surfers, even tourists have taught *honu* that humans are *safe!* That's quite a lesson to serve up to any wild creature.

Intentionally or not, Hawai'i has tamed her *honu*. It's really that simple. The challenge now is to make people aware of the enormous responsibility that comes with that taming—and to live up to it.

We leave the last words to a wise fox. When the time comes for the Little Prince to leave, the fox cries as he shares his simple secret with his new friend. As we have learned from our own experiences underwater with *honu*, taming comes with an exacting price.

"Men have forgotten this truth," said the fox. "But you must not forget it. You become responsible, forever, for what you have tamed."

Books recommended for further reading

Fire in the Turtle House: The Green Sea Turtle and the Fate of the Ocean, by Osha Gray Davidson (Public Affairs, 2001), is an excellent description of the search for the cause of fibropapilloma tumors, with special emphasis on the *honu.*

We highly recommend Archie Carr's *So Excellent a Fishe: A Natural History of Sea Turtles* (Natural History Press, 1967) and *The Windward Road: Adventures of a Naturalist on Remote Caribbean Shores,* (Knopf, 1956). Carr was not just the father of sea turtle research, he was also unsurpassed as a writer.

Another wonderful book about sea turtles in general is *Time of the Turtle,* by Jack Rudloe (Knopf: distributed by Random House, 1979).

Sea Turtles of the World, by Doug Perrine (Voyageur Press, 2003); *Sea Turtles,* by Jeff Ripple (Voyageur Press, 1996); and *Sea Turtles: An Ecological Guide* by David Gulko and Karen Eckert (Mutual Publishing, 2004) are all good general guides to sea turtle biology for lay people.

Biology of Sea Turtles, Volume I, edited by Peter Lutz and Jack Musick, and *Volume II,* edited by Peter Lutz, Jack Musick, and Jeanette Wyneken (CRC, 1996 and 2002, respectively) are excellent reference works for the serious reader.

Sea Turtles of Hawaiʻi, by Patrick Ching (University of Hawaiʻi Press, 2001) is a terrific introduction to *honu* for readers of all ages. For the youngest readers, Tammy Yee's books about Baby *Honu* (Island Heritage Publishing) are unique. We especially recommend *Baby Honu's Incredible Journey.*

Web sites recommended for further reading

Turtle Trax, www.turtles.org, is our Web site. There you can find more information about the *honu* mentioned in the book, their environment, and their struggle with fibropapilloma tumors.

Pacific Islands Fisheries Science Center Marine Turtle Research Program, www.pifsc.noaa.gov/psd/mtrp/, provides links to a wealth of scientific information about Hawaiian turtles. In particular, the 1979 George Balazs paper *Synopsis of biological data on the green turtle in the Hawaiian Islands* contains valuable information for the serious student of *honu.*

Malama na Honu Foundation at Laniakea Beach, www.malamanahonu. org, is devoted to the basking *honu* on the North Shore of Oahu.

Hawaiʻi Preparatory Academy/NOAA Sea Turtle Research and Conservation Program, www./hpa.edu/turtle.html, is the Web site for a unique collaboration that provides high school students the opportunity to work with and help the *honu.*

seaturtle.org, www.seaturtle.org, is easily the most comprehensive collection of online sea turtles resources.

It is the nature of the Web that some of these sites may move or disappear completely in the future. If these URLs fail and Google can't find their new homes, we suggest searching the *Internet Archive* at www.archive.org.

About the Authors

Peter Bennett has been documenting software for large-scale computer systems for over 25 years. He learned to dive in Hawai'i in 1987. Ursula Keuper-Bennett is a retired middle school teacher who learned to dive in 1972. They encountered their first sea turtle in 1988 while diving on the reefs of Honokōwai, Maui, as visitors from Canada. Since then, they have returned to spend every summer at Honokōwai, diving with and learning about the *honu* and paying special attention to the tumors that plague the turtles. Using photographs and videotape, they developed a system for recognizing individuals. They identified over 750 *honu,* many of them permanent residents.

In 1995, they developed *Turtle Trax,* the first Web site devoted to sea turtles. With their images, they worked to raise awareness of the tumor problem, showing how the disease affected specific *honu* over the years. Their photos have appeared in numerous publications, including the covers of *Science News* and *AWI Quarterly.* This led to an invitation to speak to the 1997 Annual Symposium on Sea Turtle Biology and Conservation, and eventually a series of presentations and posters at subsequent Symposia. They were the first to document natural regression of the tumors, and they pioneered the use of underwater photographs and videotape to study sea turtles. In recognition of the contributions made to sea turtle conservation and understanding the tumor disease, in 2000 they were nominated to the Marine Turtle Specialist Group of the Species Survival Commission of the International Union for the Conservation of Nature.

Peter and Ursula continue to spend each summer diving with the *honu* of Honōkowai.